THE
Planetary Exodus

by Daniel E. Almonz

It will happen, the revealing of an evil world pharaoh's identity before the rapture escape of Messiah's followers from earth.

The Planetary Exodus
Copyright ©2003 by Daniel E. Almonz

Printed in the United States of America

ISBN 1-594671-26-5

All Rights Reserved. No part of this publication may be reproduced, stored in a retrieval system, or transmitted in any form by any means–electronic, mechanical, digital, photocopy, recording, or any other–except for brief quotations in printed reviews, without the prior permission of the author.

All Scriptures quoted are from the King James Version. Interpretational explanations were made by the author.

Xulon Press
www.XulonPress.com

Xulon Press books are available in bookstores everywhere, and on the Web at www.XulonPress.com.

Introduction – vii

Chapter One – 1
The Globalist's CEO
The Plan for Global Government
The Roots of Global Government
The C.E.O of the Global Government
The World Pharaoh's Rise to Power • Conclusion

Chapter Two – 14
Does History Repeat Itself?
Historical Repetitions? • A Replay of History • Conclusion

Chapter Three – 27
Counterfeit Magic
Did Elijah Come Before Messiah? • Elijah and Israel's Messiah • The Forerunner of the False Messiah
Lying Wonders • The Pharaoh's Magicians
Conclusion

Chapter Four – 47
The Demonic Door
The Thessalonian Briefing • The What and the Who
Restrainer Hermeneutics
The Restrainer • Conclusion

Chapter Five – 70
Winged Eagles
The Times of the Gentiles
The Historical Scope of the Olivet Discourse
Early Church Fathers Interpret the Eagles and the Carcass
Conclusion

Chapter Six – 96
The Messiah and the Temple
Who Rebuilds the Temple?
The Temple and the Temptation
Two Signs of the Day of the Lord
The Apostasy and the Revealing • Conclusion

Chapter Seven – 136
The Shofar Blows
Jesus and the Trumpet
Did Messiah Mention the Rapture?
The Response on Earth to the Rapture
The Thessalonian Parallel • Jesus and the Rapture?

Chapter Eight – 155
The Preservation
The Future Exodus from Earth
Plagues Precede the Exodus
Future plagues Parallel the Plagues on Egypt

Chapter Nine – 181
No Intervening Events Required?
The Power Trip • An Expanding Prophecy
Concentric Circles • Evangelism and the Rapture

Chapter Ten – 202
The Gift of Prophecy
Future Knowledge
Spiritual Future Knowledge
New Covenant Glimpses of the Future
Have Spiritual Gifts Stopped?
The Gift of Prophecy and Foretelling
Mutually Contradictory

Chapter Eleven – 238
Angel of Hope
An Angel Speaks • Did the Apostle have an Imminent Hope? • Is Divine Delay Normal?

Chapter Twelve – 256
Stretch Forth Thy Hands
A Divine Perspective and the Rapture
First Century Imminency • Scattered Sheep

Chapter Thirteen – 285
The Heavenly Synagogue Meeting
The Day • The Signed but Sign-less Rapture
An Intended Pun? • Greek and "Synagogue"
The Observable Day

Chapter Fourteen – 305
The Non-Blessed Hope?
The Greek Word for Tribulation
Non-Tribulation Tribulation
The Nature of the Blessedness
Watching for Signs, not Messiah?
Not Appointed to Wrath • Conclusion

Chapter Fifteen – 328
Sign Language
The Family Feud for Imminency
The Biblical Basis for Imminency
Scripture Interpreting Scripture
Developing Sign Language Skills • Conclusion

A Word – 355

References – 358

Introduction

This book arose out of a personal search for the Biblical answers to an event known as the rapture, the sudden disappearance from earth of Messiah's followers. The specific issue concerning that event which generated my interest was the chronological question, "Is the rapture a pre-tribulational, mid-tribulational, or post-tribulational event?" I did not intend to write this book, but after having investigated the major perspectives by various authors who held the major views, evidence was discovered which led me to a conviction that I had found details a vast majority of apocalyptic interpreters had either previously overlooked, or details which they had discounted through apparently poor judgment or bias.

Some of the conclusions I had come to were arrived at independently (or by aid of the Holy Spirit). After having developed some interpretations independently, I soon discovered that I was not the only one to arrive at some of the views I had settled upon.

My research into this issue actually began while in Bible college, almost three decades ago, when as a student I chose to investigate the different views about the rapture for a theology class paper. My research at that time was handicapped by a restrictive deadline (I had to submit my paper before the semester ended), and by the limited resources available to me at that moment.

Initially, I settled upon the pre-tribulation rapture theory as the "correct" interpretation, because of all the materials I had examined, that view had the most convincing and persuasive literature supporting it. I was particularly influenced by the work of a major scholar for the pre-tribulational rapture view during the previous century who is now-deceased .

Several years later the whole issue reopened, and I decided to once again investigate the different chronological positions for the Biblical event currently called the rapture. I had been dissatisfied with the limited materials I had uncovered at my first investigation of the issue.

It was approximately two decades ago that I had decided to reopen this investigation. I was not handicapped by an artificial deadline this time around. With a little more money, and much more time, I seriously set out to again research the chronological timeline for the Biblical rapture. Since then I have examined hundreds of writings. My research has spanned the centuries of church history. I have accumulated writings by authors on every major view.

After having arrived at some independent interpretations, I soon discovered many other writers, some spanning several centuries, who had arrived at some of the same conclusions long before I had come to them. I share some of those findings in this book.

I should state that I believe the Bible is infallible according to the original autographs (the original manuscripts written by the prophets, apostles, and their secretaries). Since those original manuscript editions of the Bible no longer exist, we are left with fallible copies. It is the task of believers to examine textual differences to recapture the infallible

Biblical text. Despite some of the textual variations, I believe that most of the text we possess today is actual Scripture. The Bible is the ultimate and final authority, and for that reason, I reference Biblical texts for authoritative support. Human interpretation and judgment may be in error, but the Lord is never wrong.

I believe the Bible is the divine Word of the Lord. I believe in the miracles of the Bible, and I believe Jesus is the same yesterday, today, and forever. God is still performing miracles.

I believe in a literal thousand year future millenial kingdom on this planet.

I believe that Jesus is divine, and that He is Almighty God and Lord. I believe God has revealed Himself in three persons (the Trinity), and I believe in the future bodily return of Jesus to this earth.

As for the rapture, Biblical evidence suggests that it will one day be widely scoffed at and mocked as a doctrine (II Peter 3:3-4).

Many years ago, as a student in Bible college, I was practicing with my newly found knowledge of the Greek language. I happened to examine a classic Biblical rapture passage in the Greek. After translating the passage, I was completely convinced that, according to the Greek, there would be a rapture. My faith about that issue has never been shaken.

Unfortunately, I have met a number of believers through the years who did not share my same certainty about a future rapture. Of course, most didn't know anything about the Greek language. Maybe learning a little of the language of the New Covenant would help many of them resolve their uncertainty about that Biblically predicted future event, the rapture. The Greek language has the ability to quite remarkably convey very precise ideas.

The author.

Chapter One

The Globalist's CEO

In case you haven't noticed it lately, there is growing, massive political sentiment for worldwide, global government. It is possible you may have missed recognizing this political movement, but it is likely that some readers of this book may already be sympathetic with the whole concept.

This trend in worldwide opinion towards globalism can be observed in recent times in the move to gain a general consensus among United Nations leaders during attempts to eradicate international terror threats. While the interests of many countries are at odds with those of the United States, the desire for a general consensus among United Nations leaders seemed to be a major factor of concern for some citizens considering the issue of dealing with terrorism. An Islamic belief lies at the foundation of values for perhaps the majority of U.N. leaders. In contrast, the United States has historically and generally held to a Biblical perspective in matters of law and politics. Although this Biblical foundation for American values has been recently eroding very rapidly, gaining a consensus of opinion from an Islamic majority of United Nations leaders may produce some unanticipated results in world politics, especially as they relate to Israel.

For most people, the political movement toward a world

government may not be an event that is quite all that shocking. If anything, many may rather possess the attitude, "Why does it take so long for world government to get here?" What may or may not be surprising, or even astounding for some, is the revelation that this world government movement has already been predicted by holy men centuries ago in the Bible.

Perhaps many see attempts at globalization as the plans of men to achieve peace and unity. In reality, world globalization efforts have been not only forecast in the Scriptures, but those efforts are Biblically predicted to bring about a counterfeit unity of world nations resulting, according to the Bible, in the worldwide rejection of the Creator of our Universe.

World government and globalization efforts are ostensibly an attempt to induce international peace through government, but undermining this current effort of international leaders is none other than the serpent of old, the archenemy of man, the one sometimes in the Scriptures referred to as the "dragon," who is also called the "Adversary," or "Devil." A Satanic conspiracy undermining world unity is something holy prophets predicted in the Scriptures centuries ago.

In our own day and time we have seen the unification of Europe through the EU (European Union). The United Nations itself is, by its very own name, an attempt at world union. Prior to the United Nations, the League of Nations had attempted to bring international unity.

Perhaps we should not question the fact that many involved in the work of the European Union and the United Nations are making sincere efforts for real global peace.Peace itself is not a bad goal, if defined correctly. The problem is that the most noble efforts of man can be undermined by spiritual forces far greater and more powerful than any we might even suspect.

Perhaps the unwitting result of these peace efforts, for many, will be the empowerment of an individual of such a vile and wicked character, we could unashamedly call him the earth's future world pharaoh. He will be, in several ways, a stunning replay of the ancient pharaoh who presided during Israel's exodus from Egypt.

The Plan for Global Government
Is the plan for international unity itself flawed? Is there a fundamental problem in the very concept of a globalized world government?

The plan for world union sounds very practical and feasible, on the surface. The idea of stopping war by creation of a one-world government is superficially valid. There is just, perhaps, a possible little flaw in the plan: where can you get a human being with the needed virtue and character to run the thing?

The truth of the old adage, "Power corrupts, and absolute power corrupts absolutely," is not about to evaporate with the advent of a globalized government. In fact, the very self-exaltation of the predicted world ruler will, in and of itself, demonstrate to many the truth of that wise proverb more eloquently than could any oratory. The character of that future world ruler will secretly be of such a nature that, were it made visible, there would be little wonder as to why the Scriptures call that future monarch the "man of sin" (II Thess. 2:3). He will be thoroughly corrupt while possessing a power and authority unparalleled by any other government leader in human history.

The Roots of Global Government
How did the move toward international governmental unification begin? Whose brain-child was it to start with?

The answers to these questions are rather shocking, because the story begins several thousand years ago, during the era of a man named Nimrod who founded a kingdom in the area we know today as Babylon, or Iraq.

Nimrod's name means "rebellious" or "let us rebel." In Genesis 10:8 we learn that Nimrod "began to be a mighty one in the earth." We also learn that the beginning of his kingdom was in the land of Shinar (Gen. 10:10). It was in Shinar that the tower of Babel began as a building project for the purpose of bringing social and geographical unity. It is this kingdom of Nimrod's which was the embryonic seed for the nation which has become synonymous with the idea of sin throughout the centuries. That nation is known today by the name of Babylon.

In Genesis (1:28, 9:1, 7) the Lord's instructions to humanity were to populate the earth. At Babel, in direct rebellion and in opposition to the Lord's directive, these people proposed a unification strategy to prevent themselves from being scattered. The building project included a city and also a tower that would be thrust into the heavens. Their purpose was to make themselves renowned, and to prevent a disintegration of their population from a particular geographical location.

That the Lord was displeased at this opposition to His own plan for men to colonize this planet can be seen in His reaction.

5 And the LORD came down to see the city and the tower, which the children of men builded.

6 And the LORD said, Behold, the people is one, and they have all one language; and this they begin to do: and now nothing will be restrained from them, which they have imagined to do.

7 Go to, let us go down, and there confound their language, that they may not understand one another's speech.

8 So the LORD scattered them abroad from thence upon the face of all the earth: and they left off to build the city.

9 Therefore is the name of it called Babel; because the LORD did there confound the language of all the earth: and from thence did the LORD scatter them abroad upon the face of all the earth.
(Genesis 11:5-9)

While these men at Babel were developing a strategy for social union, the Lord deliberately began to oppose their plan and introduced an element to deter their unity. The Lord disrupted their language so that they couldn't communicate. The result was that the Almighty's purpose was fulfilled anyway. The people at Shinar were scattered to populate the world just as the Creator had originally directed. These people became obedient to the Lord's plan despite their unwillingness.

This experience at Babel illustrates the concept that unification and unity do not always correspond to the plans of the Grand Designer. Sure, the word unity has a great ring to it, but is it necessarily the primary will of the Sovereign Lord? Clearly at Babel, the movement towards unification contradicted the divine Creator's purpose and plan for world colonization. There at Shinar, the seeds of rebellion were sown, and Babylon became a nation that has opposed Israel and the people of the Lord throughout its history.

To illustrate the repugnance the Almighty Lord had towards this activity at Babel, we only have to examine the last book of the Bible and note that the kingdom of the prophesied coming world pharaoh is identified by the name of that rebellious kingdom at Shinar. The future world dictator's headquarters, is at the very least, symbolically, and very likely literally, given the name "Babylon" (Rev. 17:5).

Biblical students and interpreters are debating the meaning of the Babylon found in the concluding book of the New Covenant Scriptures. The roots of Babylon's history necessarily extend back to the plains of Shinar and the kingdom founded by Nimrod. Forever afterward, Babylon has throughout history been known as a kingdom in opposition to the people of the Lord. It is this evil spiritual heritage which the holy prophets of Scripture paint the final despotic world ruler as governing and directing.

In the book of Revelation, the Apocalypse, Babylon has become symbolic of the empire of Satan. It has become the kingdom in opposition to the kingdom of the Lord.

The strategy used at Shinar for maintaining a cohesive, unified society is rather fascinating. That strategy included building projects that were both commercial and religious.

> *4 And they said, Go to, let us build us a city and a tower, whose top may reach unto heaven; and let us*

> *make us a name, lest we be scattered abroad upon the face of the whole earth. (Genesis 11:4)*

First they would build a city. Then they would build a tower.

The city has commercial, monetary, and political connotations. The tower is believed by archeologists to have been a ziggurat, which probably had intensely religious significance. Ziggurats were used for pagan religious practices. The remains of the ziggurats we have today may have been inspired by the earliest tower at Babel. Some of these ziggurats reached as high as six levels, units, or stages. Other ziggurats reached up into a seventh level. At the top level of these towers, there existed a worship center which housed or enshrined an image of the pagan god for whom the ziggurat was erected. At least one ziggurat has been associated with the Zodiac, and some believe that Babel was the origin of astrology. So at the heart of this rebellious kingdom at Shinar, there very likely existed a counterfeit religion which may have been the central factor used to promote a plan for social unity. Counterfeit religion may have been the key strategy used to prevent the people from dispersing. The whole plan was in direct opposition to the purpose the Lord had for man.

This early start of the kingdom of Babylon has tremendously important significance, for it is a prototype of the kingdom the future world dictator will govern and manage. Babylon's spiritual significance is that it is symbolically the kingdom of the Devil, so the details we see in miniature at the beginning of this kingdom at Shinar, we will see more fully developed in the final incarnation of that empire.

The early start of the kingdom of Babylon at Shinar had political, commercial, and apparently religious aspects. Many leading students of the Bible believe that the final Babylon of the future world pharaoh will have political, commercial, and religious aspects as well. At the heart of this final rebellious form of the evil empire will be a counterfeit religion.

Man is by nature religious, and the future despotic world monarch is predicted in the Scriptures to use religion as a part of his plan, just as his ancient predecessor Nimrod had apparently done at the inception of the Babylonian empire on the plains of Shinar.

The C.E.O of the Global Government

A global government needs a Chief Executive Officer to run it. Governments are not run by computers. People run governments. The Biblically predicted global government will have its Chief Officer, and the Scriptural predictions about him are so numerous that speculations about his identity rival the diversity of speculations that exist about the coming of the Messiah.

In the vision given to the Apostle John on Patmos, the last book of the New Covenant, we read about this future pharaoh.

> *3 ... and all the world wondered after the beast.*
> *4 ... and they worshipped the beast, saying, Who is like unto the beast? who is able to make war with him? (Revelation 13:3, 4)*

These passages of Scripture call the world pharaoh a "beast" (Rev. 13:1). The Greek word refers to a "wild beast."

> Man becomes "brutish" when he severs himself from God, the archetype and true ideal, in whose image

he was first made, which ideal is realized by the man Christ Jesus. Hence, the world-powers seeking their own glory, and not God's, are represented as beasts; and Nebuchadnezzar, when in self-deification he forgot that "the Most High ruleth in the kingdom of men," was driven among the beasts. In Daniel 7. there are four beasts: here the one beast expresses the sum-total of the God-opposed world-power viewed in its universal development... (JFB, n.d., Rev. 13:1)

The Greek word which signifies a "wild beast" suggests that the character of this world leader is not tamed, but in rebellion to God. He is called a beast for several possible reasons, among which is the fact that his character is so uncivilized and lacking in noble virtues, he can be considered ethically to be similar to a brutish wild animal.

These passages indicate the scope of the future world dictator's influence and power. His power or influence extends, in some sense, apparently to the whole planet, for the whole earth wonders after the beast, and most of earth's population worships him at a particular point of his political power. He also appears to have some kind of authority over almost every language group, indicating his political authority approaches, if not includes, virtually the entire planet at the apex of his reign of terror.

> 7 ... *and power was given him over all kindreds, and tongues, and nations.*
> 8 *And all that dwell upon the earth shall worship him, whose names are not written in the book of life of the Lamb slain from the foundation of the world. (Revelation 13:7, 8)*

This global dictator will arise to power and exercise control politically, religiously, and commercially, but his real nature is that of a beast. We have already examined Biblical passages which indicate the vile ruler's political control, but the wicked world pharaoh will also become the center of a false religious system, for he has as an ally a powerful religious leader of great influence who will cause the world to reverence and glorify some sort of moving, visual reproduction or representation of the tyrant. This allied religious leader also has the same evil character of the dictator, for he is called a beast as well.

> *11 And I beheld another beast coming up out of the earth; and he had two horns like a lamb, and he spake as a dragon.*
>
> *12 And he exerciseth all the power of the first beast before him, and causeth the earth and them which dwell therein to worship the first beast, whose deadly wound was healed.*
>
> *13 And he doeth great wonders, so that he maketh fire come down from heaven on the earth in the sight of men,*
>
> *14 And deceiveth them that dwell on the earth by the means of those miracles which he had power to do in the sight of the beast; saying to them that dwell on the earth, that they should make an image to the beast, which had the wound by a sword, and did live.*
>
> *15 And he had power to give life unto the image of the beast, that the image of the beast should both speak, and cause that as many as would not worship the image of the beast should be killed. (Revelation 13:11-15)*

The worldwide scope of this wicked pharaoh's influence will include Israel, and for a time, the headquarters for much of his work will be there in the land of the Bible (Daniel 11:45). Amazingly, he will violate the spirit of the ten commandments with his religious system which requires actual worship of some kind of visually moving "image" of himself. Whether this "image" is a video, film, robot, or holographic movie, or some other type of technology, its purpose will be to serve as an object of worship.

In addition to the worldwide political power and religious influence of this global pharaoh, he will exert commercial power over the world through his ally, the second beast.

> *16 And he causeth all, both small and great, rich and poor, free and bond, to receive a mark in their right hand, or in their foreheads:*
>
> *17 And that no man might buy or sell, save he that had the mark, or the name of the beast, or the number of his name. (Revelation 13:16, 17)*

This future world dictator will make the tyrants of history look like grade school children in comparison.

The World Pharaoh's Rise to Power

One may question how an evil reprobate like the coming global ruler could rise to such power while possessing such a vile and murderous character. The answer to that question is long and involved, but there are some short answers that help us to realize its reality.

The coming world monarch will apparently gain political power by masquerading as a "good guy." In reality, the Scripture in II Thessalonian 2:3 calls him the "son of perdition," a term used to describe the betrayer of Jesus, Judas Iscariot (John 17:12). Like Judas, the wicked world Pharaoh will conceal the real nature of his character.

> *14 And no marvel; for Satan himself is transformed into an angel of light.*
>
> *15 Therefore it is no great thing if his ministers also be transformed as the ministers of righteousness; whose end shall be according to their works. (II Corinthians 11:14, 15)*

The masquerade will be far greater than just the concealment of his evil and vile nature. He will apparently make the claim that he is Israel's Messiah, and will successfully seduce that nation into believing him. The genuine Messiah made that specific prediction almost 2,000 years ago.

> *43 I am come in my Father's name, and ye receive me not: if another shall come in his own name, him ye will receive. (John 5:43)*

It is very likely Israel will be deceived into accepting this counterfeit as her expected King. The future world ruler, by concealing the evil and wicked nature of his character, will apparently fit the stereotype Israel has historically held about her Biblically predicted Messiah. This will result in initial acceptance of the fraudulent counterfeit prince by the nation of Israel.

Conclusion
A world government makes sense, if you can find the right C.E.O. to run it. Few would want a man like Hitler, Mussolini, or Stalin to run a global government. This is the whole problem with the idea for world government: finding

the right individual with proper character to manage it.

The Holy Scriptures predicted centuries ago that Satan would promote his own candidate for the position of ultimate world leader. That is a position which rightly belongs to the legitimate Messiah, Jesus, who fulfilled over 300 Messianic predictions made by the Biblical prophets of the past. Jesus was rejected as Messiah by the leaders of Israel in His day when He came almost 2,000 years ago, but He predicted that someone who would come in their own name would, in the future, be widely accepted as the Messiah. This was very likely a prediction about the future counterfeit Messiah, the future world pharaoh.

The good news is that Jesus is going to return. The bad news is that before Jesus actually returns to earth to assume His rightful control of this world, a counterfeit, or false, Messiah of despicable character will first gain unparalleled political power over this planet. Other good news consists in the fact that the Scriptures predict that Jesus will snatch away His followers in a planetary exodus, after the identity of the evil world pharaoh is revealed. The Bible predicts a rescue which will save the real Messiah's followers from the horrible tyranny of the future evil pharaoh's apparently global government.

Future chapters will examine some of the Biblical predictions about the coming world pharaoh. Biblical predictions about the future rapture rescue will also be examined..

Chapter Two

Does History Repeat Itself?

Many eons ago, while I was in Junior High School, there was a display in a glass case that had been set up in the school hallway by the school's Administration. Inside the glass case were a picture and a large printed sign. The sign made comparisons about two former United States Presidents. The printed sign was enclosed in the glass case, along with a picture of the late President John F. Kennedy.

The glass enclosed display was set up, evidently, to highlight some amazing similarities between two assassinated Presidents: John F. Kennedy, and Abraham Lincoln. I do not recall the entire list of similarities, but there were a few comparisons which have remained with me to this day.

The display pointed out that both Lincoln and Kennedy were Presidents of the United States. It further stated that both were shot in the head, apparently by lone assassins. Both were shot while their wives were with them. Both died of bullet wounds to the head. The sign pointed out that both had Vice Presidents named Johnson. In addition, both were succeeded in office by their Vice Presidents, so that both successors were named "Johnson." These are only a few of the items in the list.

After reading the sign in that glass enclosed display, I developed an eerie feeling, a feeling almost as if I had personally taken a tourist's trip to the Twilight Zone. The sign raised a question which it never philosophically solved. The headline on the sign asked, "Does History Repeat Itself?" Apparently, the question was supposedly answered by the list of similarities about both assassinated Presidents.

In retrospect, one may conclude that Johnson is a very common name. One could conclude that two Presidents having as their Vice Presidents men by the name of Johnson, while unusual, was coincidental, but somewhat short of the miraculous. After all, the name Johnson, from a simple perusal of the phone book, is not all that unusual. While the similarities between Presidents Lincoln and Kennedy do have some striking parallels, the question on the display still deserves an answer. Does history repeat itself?

The sign in that Junior High School seemed to suggest some eerie metaphysical reality which had not yet been recognized or defined by scientific observation or analysis. The question still remains: does history repeat itself?

Historical Repetitions?
We can define "history" as a transmitted written record of past events. If we define history in this sense, we are not solving the question, "Does history repeat itself?" The question is not asking whether duplicate accounts or similar records of transmitted past events exist. The question really being asked is, "Do events in human experience exhibit a pattern of recognizable repetition?"

From what aspect or perspective should we answer this question? Most human beings recognize a certain regularity in human events. The law of gravity is quite predictable, and this pattern of regularity occurs quite understandably in nature without surprise in human experience. In addition, there are other recognized laws of physics which occur quite punctually in human experience, so much so, that volumes of textbooks have been written just on the subject of physics alone. We know that rainbows occur regularly in human experience, and that they are the products of light being refracted through water.

"Does history repeat itself?" In examining this question which appeared on that sign in my Junior High School, we can determine that what is really being asked is, "Are there historical patterns that repeat themselves in human experience?"

If we examine nature historically, we recognize a pattern in the seasons. There are cycles of summer, fall, winter and spring. We see a cycle in individual human histories as well: babyhood, childhood, teen-hood, adulthood, parenthood. These cycles are repeated in individual lives. History does repeat itself through physical laws like gravity, and the rotation of the earth. We also observe chronological cycles in the development of animals and humans from infancy to death.

"Does history repeat itself?" When one examines that question, one somehow gets the idea that the question really being asked goes something like this: is there some mystical element at work in history which results in repetitions which transcend the laws of physics and nature?

Are there elements in history which result in the repetition of events which transcend the laws of physics? The answer to this question must be a resounding, "Yes!" We can all point to moral laws which exist in human experience. There are the patterns which repeat themselves in human history when obedience occurs, when moral laws are obeyed. Psychologists and psychiatrists point to patterns of social behavior that recur when rebellion is practiced. Whole categories of psychological names have been defined to identify people with social or mental problems, because these patterns are recognized and repeated, not just in one individual's life, but in the lives of multitudes of people when specific moral laws are confronted and rejected. History exhibits the repetition that occurs in the lives of individuals, families, and nations when certain social laws are violated and trampled.

There are moral laws which repeat themselves in the history of human psychology with such regularity that psychiatrists and counselors have actually developed categories of names for defining recognizable patterns in human behavior.

There also exist spiritual laws. According to the Scriptures, there are divine spiritual principles which exercise control over human affairs and events. Consider the following Biblical spiritual principle found in the letter of the Apostle Paul to the Galatians.

> *7   Be not deceived; God is not mocked: for whatsoever a man soweth, that shall he also reap.*
>
> *8   For he that soweth to his flesh shall of the flesh reap corruption; but he that soweth to the Spirit shall of the Spirit reap life everlasting. (Galatians 6:7, 8)*

Here we read of a spiritual law which is compared to farming. It is the law of planting and reproduction. In biology it is included under the principle that "like begets like." This law of agriculture is used to suggest that it also applies to a spiritual type of "farming." This section of Scripture also states that moral acts are worthwhile.

> *9 And let us not be weary in well doing: for in due season we shall reap, if we faint not.*
>
> *10 As we have therefore opportunity, let us do good unto all men, especially unto them who are of the household of faith. (Galatians 6:9, 10)*

Yes, history repeats itself regularly. The disobedience practiced by individuals often generates events that result in their destruction. Disciplined habits of obedience often result in predictable victories, triumphs, and success.

The existence of spiritual and moral laws allows comparison between the character of one rebellious person and the character of another. The pattern of events that characterize the rebellion of an individual exhibit remarkable similarities and parallels to the pattern of events which accompany other individuals who are in rebellion. The existence of moral, psychological, physical, and spiritual laws enable the observer to recognize patterns which predictably repeat themselves in the events which accompany human behavior. It doesn't take a nuclear physicist to predict that someone who makes a habit of robbing stores will likely end up in prison.

It is just because of these different laws of nature, laws of psychology, laws of morality, and laws of spiritual existence that we can make analogies between persons and events in history and extract astounding parallels about them. This, in fact, is done regularly in the Scriptures, and a whole system of interpretation in the Biblical writings is based on this concept. That system of interpretation has been given the name "typology." "Typology" identifies parallels between individuals, behaviors, and events. These parallels exist because of the existence of moral, spiritual, and psychological laws.

While some Bible students suggest that only Biblically sanctioned "types" are acceptable, the very existence of spiritual and moral laws suggests there must exist numerous parallels between people and events in the Scriptures which have never been mentioned by any of the Biblical writers. Indeed, the Rabbis have often made comparisons about Biblical persons which have never been officially made by any writer of Scripture.

The system of Biblical interpretation called "typology" is not necessarily some mystical magic mirror, but it is a developed ability and skill allowing some to observe similarities and parallels based on moral behaviors and divine spiritual laws which have had continuity throughout human history. It is because of these spiritual and moral laws and principles that one can make comparisons between Israel's King David, and David's greatest offspring, the Messiah. Events and situations in King David's life have prophetic and spiritual parallels in the Biblically foretold ministry of the Messiah. David is a prototype of his greatest descendant, the Messiah. We find in David's life a foreshadowing of the ministry of the predicted future righteous King of Israel, the Messiah.

A Replay of History
What of the future wicked monarch who will control the world? Are there Biblical parallels, foreshadowings, or

prototypes of this evil king? The answer is yes.

During Israel's exodus from Egypt we discover that the Pharaoh who presided provides us with a dramatic picture of the future world ruler. The Scriptures actually call the future evil ruler, "the anti-Messiah" ("the antichrist").

The Biblical writings which compose the New Covenant, or New Testament, were composed in, or have been preserved primarily in, the Greek language. Biblical scholars in examining these Greek writings will often speak of "semitisms." Semitisms are Hebrew or Aramaic language patterns which have been preserved or transmitted into the Greek language of the New Covenant, and are identifiable because the original writers and authors were semitic speaking people (they spoke Aramaic and/or Hebrew). In composing the letters and writings of the New Covenant Scriptures, the Jewish writers and authors thought in a semitic mode, but transmitted their concepts with a non-semitic language, Greek. For this reason, the Jewish nature of the New Covenant Scriptures is sometimes hidden because the language of transmission is a non-semitic language.

The Greek language, which was proliferated by Alexander the Great when he conquered the New Testament region of the earth, became the common language of commerce in the New Testament world. For practical reasons, the New Covenant Scriptures were preserved in the language of the common man. The Hebrew word for "Messiah" is translated as "Christos" in the Greek. In English the world has come down to us as "Christ." "Messiah" means "the Anointed One." "Christ" means "the Anointed One." When the New Covenant writings refer to the future world ruler as the "anti-Christ," they are really referring to him as the "anti-Messiah." The future world ruler will be a pseudo-Messiah. He will be a "false Messiah," a counterfeit of the real Messianic King predicted in the Hebrew Bible.

The anti-Christ, or pseudo-Messiah, will attempt to "steal the show." His world rule will be a pathetic imitation or attempted replacement for the real Messiah's prophesied government, one predicted in the following passage by Isaiah.

> 6 For unto us a child is born, unto us a son is given: and the government shall be upon his shoulder: and his name shall be called Wonderful, Counsellor, The mighty God, The everlasting Father, The Prince of Peace.
>
> 7 Of the increase of his government and peace there shall be no end, upon the throne of David, and upon his kingdom, to order it, and to establish it with judgment and with justice from henceforth even forever. The zeal of the LORD of hosts will perform this. *(Isaiah 9:6, 7)*

While the anti-Messiah is not necessarily called a "pharaoh," in the Bible, the pharaoh of Israel's exodus provides us with one of the most detailed prototypes of the future anti-Messiah in all of Scripture. Let us examine some of these parallels.

First, a parallel exists between the pharaoh of the Egyptian exodus and the future pseudo-Messiah in that the Egyptian pharaoh spoke blasphemously against Israel's deity (Pink, 1923, pp. 226-227).

> 2 And Pharaoh said, Who is the LORD, that I should obey his voice to let Israel go? I know not the LORD, neither will I let Israel go. *(Exodus 5:2)*

While we may question whether the above language was blasphemous, it should be recalled that the Egyptian pharaoh was evidently considered to be a deity himself. For Pharaoh as a god to state that he had no knowledge of Israel's deity was, at the very least, an insult.

The anti-Messiah is pictured in the New Covenant as one who also speaks blasphemously against Israel's Lord.

> *5 And there was given unto him a mouth speaking great things and blasphemies; and power was given unto him to continue forty and two months.*
>
> *6 And he opened his mouth in blasphemy against God, to blaspheme his name, and his tabernacle, and them that dwell in heaven. (Revelation 13:5-6)*

Secondly, another parallel exists between the anti-Messiah and the pharaoh of the exodus. In Egypt, the pharaoh endured the supernatural plagues and signs at the hands of the Lord's two witnesses, Moses and Aaron. In like manner, the future world pharaoh is pictured in the Apocalypse as one who observes the signs of two miracle working witnesses of the Lord.

> *3 And I will give power unto my two witnesses, and they shall prophesy a thousand two hundred and three score days, clothed in sackcloth.*
>
> *6 These have power to shut heaven, that it rain not in the days of their prophecy: and have power over waters to turn them to blood, and to smite the earth with all plagues, as often as they will.*
> *(Revelation 11: 3, 6)*

Thirdly, the supernatural plague of water turning to a red color so that it could be called "blood" was witnessed by the pharaoh of the exodus, and a parallel demonstration is witnessed by the future world pharaoh in Revelation 11:6, quoted above.

Fourth, another similarity between the future world pharaoh and the pharaoh of the exodus can be observed in the self-declaration of godhood by the false Messiah.

4 Who opposeth and exalteth himself above all that is called God, or that is worshipped; so that he as God sitteth in the temple of God, shewing himself that he is God. (II Thessalonians 2:4)

History reveals to us that the Egyptian pharaohs were considered divine. This megalomania may account for the hardness of heart that Pharaoh exhibited at Israel's Egyptian exodus. When Pharaoh declared, "I know not the Lord (of Israel), neither will I let Israel go," he may have been claiming that, as a divine god himself who was in a class with the gods of Egypt, he had no acquaintance with the deity from Israel. He may have been speaking sarcastically, ridiculing the idea of a real God, or he may have deluded himself into thinking he really was a deity who was of such great significance and power, that if the God of Israel had any merit, he would have already known about Him. The claim of deity has been a feature of many political rulers like some of Rome's Caesars, some of Japan's Emperors, and by Alexander the Great. The false Messiah will make the same blasphemous claim of divinity that has characterized some of the most powerful rulers of history.

Fifth, just as the pharaoh of Israel's exodus used magicians to counter the supernatural signs of Moses and Aaron, so too, the anti-Messiah will have his false prophet who performs lying wonders, perhaps to refute the two witnesses who smote the earth with plagues, whom he had killed.

> *13 And he doeth great wonders, so that he maketh fire come down from heaven on the earth in the sight of men,*
>
> *14 And deceiveth them that dwell on the earth by the means of those miracles which he had power to do in the sight of the beast; saying to them that dwell on the earth, that they should make an image to the beast... (Revelation 13:13, 14)*

There is a sixth parallel between the pharaoh of the exodus and the future world pharaoh. The Pharaoh of ancient Egypt broke his promise to the Israelites of the exodus.

> *8   Then Pharaoh called for Moses and Aaron, and said, Intreat the LORD, that he may take away the frogs from me, and from my people; and I will let the people go, that they may do sacrifice unto the LORD.*
>
> *12 And Moses and Aaron went out from Pharaoh: and Moses cried unto the LORD because of the frogs which he had brought against Pharaoh.*
>
> *13 And the LORD did according to the word of Moses; and the frogs died out of the houses, out of the villages, and out of the fields.*
>
> *15 But when Pharaoh saw that there was respite, he hardened his heart, and hearkened not unto them; as the LORD had said. (Exodus 8:8, 12, 13, 15)*

Just as the pharaoh of Egypt broke his promise to the Israelites, so the future world ruler is predicted by Scripture to break his treaty with Israel

> *27 And he shall confirm the covenant with many for one week: and in the midst of the week he shall cause the sacrifice and the oblation to cease, and for the overspreading of abominations he shall make it desolate, even until the consummation, and that determined shall be poured upon the desolate. (Daniel 9:27)*

Here we are given a picture of the future pharaoh making a seven-year treaty with Israel. In the midst of the seven years, the counterfeit Messiah breaks the covenant, not keeping his word, just as the pharaoh of Egypt broke his word to Moses and Aaron (Pink, 1992, pp. 226-227). These numerous parallels between the pharaoh of the Egyptian exodus and the future world pharaoh illustrate the repetitions which exist in history. Through past events we can sometimes see prototypes and foreshadowings of events yet to occur. The parallels are often so remarkable one cannot help but stand in wonder and awe at the Grand Designer's ability to arrange history in such an organized manner. At times, we are able to see the future by use of the past.

There are other foreshadowings of the coming false Messiah who becomes world ruler in the Scriptures. Glimpses of the future world pharaoh can be found, for example, in the life of Nebuchadnezzar, perhaps ancient Babylon's most renowned king.

Conclusion
History does present us with the remarkable ability to see parallel happenings and features in past events. It appears

as though the Lord has arranged His universe in this manner to enable humans to see His sovereign and mighty hand in life's affairs.

This repetition in history allows human beings to learn from other people's mistakes. A whole system of Biblical interpretation known as "typology" has developed out of these remarkably observed parallels. Typology is sometimes used by some to prove a doctrinal teaching, but typology is only illustrative of a truth, it doesn't necessarily prove a supposed parallel will exist, only that a parallel can and may exist. Pushing details too far can muddy the appropriate principles which enable valid parallels to be observed. The potential error is to suppose some presuppositions about predicted events are valid because a parallel example may exist that is strikingly similar to our conclusions about those Biblically predicted future events, but opposite parallels which run contrary to our conclusions can sometimes also be drawn as well for prophesied Biblical events. One must study what the literal Biblical passages teach on an issue in order to find appropriate foreshadowings for those future events in history. Parallels to Israel's exodus from Egypt have been used because they appear to correspond with what literal Biblical passages validly teach about the future planetary exodus, the rapture.

The next chapter will examine in detail one of the parallels we have already mentioned.

Chapter Three

Counterfeit Magic

A cup of wine is placed at the table before an empty chair every year at the same time. The cup is for an expected guest, who never shows up. Every year children run to the door to see if the expected visitor has come, but he never does. Who is this mysterious person whose presence is expected?

It is at Passover that the cup of wine is placed in front of an empty chair. The guest which Jewish participants are expecting with this custom is Elijah. Why? Why is there such emphasis placed on the coming of Elijah in Judaism?

In the Hebrew Bible, the Tenakh (which is the "Old Covenant" of the Christian Bible), the prophet Malachi made a prediction concerning Elijah.

> 5   Behold, I will send you Elijah the prophet before the coming of the great and dreadful day of the LORD:
>
> 6   And he shall turn the heart of the fathers to the children, and the heart of the children to their fathers, lest I come and smite the earth with a curse.
> *(Malachi 4:5, 6)*

This prophecy by Malachi predicts the coming of Elijah at a future time in history. Some Jews believe that these Scriptures, along with Malachi 3:1, indicate that Elijah will precede the coming of the Jewish Messiah.

> *1 Behold, I will send my messenger, and he shall prepare the way before me: and the LORD, whom ye seek, shall suddenly come to his temple, even the messenger of the covenant, whom ye delight in: behold, he shall come, saith the LORD of hosts. (Malachi 3:1)*

It is mentioned in the traditions of Judaism that Elijah is to come and bring peace to the world (Cohen, 1985, p. 356).

Did Elijah Come Before Messiah?
What was Elijah like? According to the Scriptures he was a miracle worker. Let us examine a list of some miracles associated with Elijah's life.

1) He shut up the heavens so it didn't rain (I Kings 17:1).
2) He was fed by birds during a time of famine (I Kings 17:2-6).
3) He was instrumental in multiplying oil and flour during the drought for a widow and her son (I Kings 17:8-16).
4) He raised the widow's dead boy back to life (I Kings 17:17-24).
5) He called down fire from heaven onto an altar at Mt. Carmel (I Kings 18:17-39).
6) He ended the three-and-a-half year drought through prayer (I Kings 18:41-45).
7) He called down fire from heaven twice to destroy soldiers (II Kings 1:9-15).
8) He parted the waters of the Jordan (II Kings 2:6-8).
9) He supernaturally ascended to heaven during a whirl-

wind as a chariot of horses passed by (II Kings 2:1, 9-11).

The list of these supernatural miracles in the life of Elijah naturally predisposes one to believe that if Elijah precedes the Messiah, as the prophet Malachi appears to predict, then one should expect a supernatural miracle ministry to occur before the appearance of the Messiah.

The fulfillment by Jesus of over 300 Messianic prophecies from the Jewish Scriptures statistically inclines the evidence in favor of His Messiahship. The question must then be asked, "Did Elijah precede the coming of Jesus?" Was there some miracle worker who introduced Jesus by manifesting a miraculous ministry? It is reasonable to suppose a miracle working ministry should precede the Messiah, since Malachi said Elijah would come before the Messiah.

Elijah and Israel's Messiah
Despite the hundreds of Messianic Biblical predictions fulfilled by Jesus, the question still exists, "What relationship does Elijah have to Jesus?" As far as a miracle ministry introducing the coming of Jesus is concerned, the Biblical record appears to be empty. There was no miracle ministry that preceded the coming of Jesus. Does this make Jesus a fraud? Is the preacher of Nazareth a pretender? Some would say the lack of fulfillment for this supernatural miracle working ministry preceding the ministry of Jesus does make Jesus an illegitimate contender for the office of Messiah. Before we jump to conclusions, there are several observations that can be made by a careful evaluator.

First, Malachi said "Elijah" would precede the Messiah, not miracles. Secondly, Jesus, according to the teachings of the New Covenant Scriptures, will have another coming, one in which He returns later to set up His Messianic

Kingdom. There are some Bible students who emphatically believe the predictions about Elijah preceding the Messiah are prophecies which will be literally fulfilled at the Second Coming of Jesus. Certainly no Elijah-like miracle ministry preceded Jesus at His first coming. Thirdly, we can observe that Elijah did appear during the first coming of Jesus while He walked the earth, about 2,000 years ago. At the mount where the transfiguration of Jesus occurred, Elijah did appear and talked with Jesus, along with Moses.

> *1 And after six days Jesus taketh Peter, James, and John his brother, and bringeth them up into an high mountain apart,*
>
> *2 And was transfigured before them: and his face did shine as the sun, and his raiment was white as the light.*
>
> *3 And, behold, there appeared unto them Moses and Elias talking with him.*
>
> *4 Then answered Peter, and said unto Jesus, Lord, it is good for us to be here: if thou wilt, let us make here three tabernacles; one for thee, and one for Moses, and one for Elias.*
>
> *9 And as they came down from the mountain, Jesus charged them, saying, Tell the vision to no man, until the Son of man be risen again from the dead.*
> *(Matthew 17:1-4, 9)*

That Elijah did appear on the earth during the first coming of Jesus is recorded here. There is an important observation about this appearance of Elijah to Jesus. This appearance of Elijah to Jesus was a one time event. The prediction by the prophet Malachi infers that there would be some protracted ministry by Elijah, apparently to Israel.

> *6 And he shall turn the heart of the fathers to the children, and the heart of the children to their fathers, lest I come and smite the earth with a curse.*
> *(Malachi 4:6)*

Elijah's appearance at the transfiguration of Jesus was a secret appearance without any lengthy ministry. Elijah did not fulfill this Malachi prophecy at the first coming of Jesus, He spoke only with Jesus and Moses. Elijah's appearance during the transfiguration event was a secret appearance. Should we conclude that if Jesus is the Messiah, Malachi's prophecy regarding the Elijah ministry just wasn't fulfilled when Jesus came approximately 2,000 years ago, and that we should await Malachi's fulfillment at Messiah's Second Coming?

While it seems reasonable to conclude that Elijah will come literally, or in type, before Messiah's Second Coming, Jesus indicated to His disciples that there was some contemporary form of fulfillment concerning the Malachi prediction of Elijah's ministry.

> *9 And as they came down from the mountain, Jesus charged them, saying, Tell the vision to no man, until the Son of man be risen again from the dead.*

> *10 And his disciples asked him, saying, Why then say the scribes that Elias must first come?*

> *11 And Jesus answered and said unto them, Elias truly shall first come, and restore all things.*

12 But I say unto you, That Elias is come already, and they knew him not, but have done unto him whatsoever they listed. Likewise shall also the Son of man suffer of them.

13 Then the disciples understood that he spake unto them of John the Baptist. (Matthew 17:9-13)

When John the Baptist was imprisoned, he sent some questions to Jesus. After Jesus gave His reply to the messengers, He turned to the multitudes and began to speak about the Baptist. In His tribute to the Baptizer, Jesus quoted Malachi 3:1, claiming that it was a prediction fulfilled by John.

10 For this is he, of whom it is written, Behold, I send my messenger before thy face, which shall prepare thy way before thee.

11 Verily I say unto you, Among them that are born of women there hath not risen a greater than John the Baptist: notwithstanding he that is least in the kingdom of heaven is greater than he.

12 And from the days of John the Baptist until now the kingdom of heaven suffereth violence, and the violent take it by force.

13 For all the prophets and the law prophesied until John.

14 And if ye will receive it, this is Elias, which was for to come.

15 He that hath ears to hear, let him hear. (Matthew 11:10-15)

While Jesus plainly stated John did no miracles, He did intimate that John's ministry was like the ministry of Elijah, and the New Covenant Scriptures explicitly say so. In the gospel of Luke, an account is preserved concerning the father of John the Baptist who was a member of the tribe of Levi.

13 But the angel said unto him, Fear not, Zacharias: for thy prayer is heard; and thy wife Elisabeth shall bear thee a son, and thou shalt call his name John.

14 And thou shalt have joy and gladness; and many shall rejoice at his birth.

15 For he shall be great in the sight of the Lord, and shall drink neither wine nor strong drink; and he shall be filled with the Holy Ghost, even from his mother's womb.

16 And many of the children of Israel shall he turn to the Lord their God.

17 And he shall go before him in the spirit and power of Elias, to turn the hearts of the fathers to the children, and the disobedient to the wisdom of the just; to make ready a people prepared for the Lord.

18 And Zacharias said unto the angel, Whereby shall I know this? for I am an old man, and my wife well stricken in years.

19 And the angel answering said unto him, I am Gabriel, that stand in the presence of God; and am sent to speak unto thee, and to shew thee these glad tidings. (Luke 1:13-19)

These passages relate that John the Baptist's father was a priest in the Temple. While performing his duties, the Baptist's father had a vision of the angel Gabriel who alluded to the Malachi 4:6 prophecy about Elijah's ministry. Gabriel predicted that it was the child to be born, John the Baptist, who would fulfill the Elijah prophecy.

> Comparing Malachi 3:1 with Isaiah 40:3 indicated that the Elijah Messenger (John the Baptist), was preparing the way for "Yahweh" in the flesh of the Messiah. The Baptizer was heralding the way for "Yahweh" in bodily form. (JFB, n.d., Luke 1:16).

It must be noted that Gabriel did not quote the section from Malachi about, "... the great and dreadful day of the LORD." That portion of the text does not apply to the first coming of Jesus. That portion applies to Messiah's Second Coming, which is still future and which is inaugurated by the rescue of Messiah's followers from earth at the start of day of the Lord, the rapture (I Thess. 4:16-5:5). The failure of the angel Gabriel to quote the part of Malachi's prophecy about the "the great and dreadful day of the LORD" suggests that the Malachi prediction has two applications. The Malachi prophecy was partially fulfilled at Messiah's first coming by John the Baptist. The second application for Malachi's prophecy appears to be at the time of the "day of the Lord," a term the Apostle Paul applies to the rapture in I Thessalonians (Van Kampen, 1997, p. 111).

The Forerunner of the False Messiah
That John the Baptist did no miracles didn't fit the Elijah stereotype Israel had for the coming Elijah who was predicted to precede the Messiah. Perhaps this is why Jesus said of John the Baptist, "And if ye will receive it, this is Elias, which was for to come." The Elijah forerunner to the Messiah was predicted by two different passages

in the book of Malachi (3:1, 4:5-6), but the first passage does not refer to the forerunner as Elijah, but rather as a "Messenger."

It appears that the stereotype Israel had for the Messiah's forerunner will be counterfeited by the devil through the man who is popularly called the "false prophet." The man who is represented by the second beast of Revelation 13 appears to manifest a miracle ministry which may fulfill all of the expectations that Israel has hoped for with the anticipated Elijah. The false prophet will, like Elijah of old, appear to be able to call down fire from heaven (probably lightning).

We read of Elijah calling down fire from heaven on three occasions. That fire was probably lightning. First, at Mt. Carmel, Elijah used his trademark miracle in a contest with the prophets of Baal.

> *21 And Elijah came unto all the people, and said, How long halt ye between two opinions? if the LORD be God, follow him: but if Baal, then follow him. And the people answered him not a word.*
>
> *23 Let them therefore give us two bullocks; and let them choose one bullock for themselves, and cut it in pieces, and lay it on wood, and put no fire under: and I will dress the other bullock, and lay it on wood, and put no fire under:*
>
> *24 And call ye on the name of your gods, and I will call on the name of the LORD: and the God that answereth by fire, let him be God. And all the people answered and said, It is well spoken.*
>
> *26 And they took the bullock which was given them,*

and they dressed it, and called on the name of Baal from morning even until noon, saying, O Baal, hear us. But there was no voice, nor any that answered. And they leaped upon the altar which was made.

27 And it came to pass at noon, that Elijah mocked them, and said, Cry aloud: for he is a god; either he is talking, or he is pursuing, or he is in a journey, or peradventure he sleepeth, and must be awaked.

30 And Elijah said unto all the people, Come near unto me. And all the people came near unto him. And he repaired the altar of the LORD that was broken down.

32 And with the stones he built an altar in the name of the LORD: and he made a trench about the altar, as great as would contain two measures of seed.

33 And he put the wood in order, and cut the bullock in pieces, and laid him on the wood, and said, Fill four barrels with water, and pour it on the burnt sacrifice, and on the wood.

34 And he said, Do it the second time. And they did it the second time.

And he said, Do it the third time. And they did it the third time.

35 And the water ran round about the altar; and he filled the trench also with water.

36 And it came to pass at the time of the offering of the evening sacrifice, that Elijah the prophet came near, and said, LORD God of Abraham, Isaac, and of Israel, let it be known this day that thou art God in Israel, and that I am thy servant, and that I have done all these things at thy word.

37 Hear me, O LORD, hear me, that this people may know that thou art the LORD God, and that thou hast turned their heart back again.

38 Then the fire of the LORD fell, and consumed the burnt sacrifice, and the wood, and the stones, and the dust, and licked up the water that was in the trench.

39 And when all the people saw it, they fell on their faces: and they said, The LORD, he is the God; the LORD, he is the God.
(I Kings 18:21, 23-24, 26-27, 30, 32-33, 34-39)

The second occasion for Elijah's trademark miracle of fire from heaven was when fifty soldiers were sent to take Elijah back to King Ahaziah in Samaria.

9 Then the king sent unto him a captain of fifty with his fifty. And he went up to him: and, behold, he sat on the top of an hill. And he spake unto him, Thou man of God, the king hath said, Come down.

10 And Elijah answered and said to the captain of fifty, If I be a man of God, then let fire come down from heaven, and consume thee and thy fifty.

And there came down fire from heaven, and consumed him and his fifty. (II Kings 1:9-10)

The third occasion for Elijah's trademark miracle of fire

from heaven was when another fifty soldiers were sent to take Elijah back to King Ahaziah in Samaria.

11 Again also he sent unto him another captain of fifty with his fifty.

And he answered and said unto him, O man of God, thus hath the king said, Come down quickly.

12 And Elijah answered and said unto them, If I be a man of God, let fire come down from heaven, and consume thee and thy fifty. And the fire of God came down from heaven, and consumed him and his fifty.

13 And he sent again a captain of the third fifty with his fifty. And the third captain of fifty went up, and came and fell on his knees before Elijah, and besought him, and said unto him, O man of God, I pray thee, let my life, and the life of these fifty thy servants, be precious in thy sight.

14 Behold, there came fire down from heaven, and burnt up the two captains of the former fifties with their fifties: therefore let my life now be precious in thy sight.

15 And the angel of the LORD said unto Elijah, Go down with him: be not afraid of him. And he arose, and went down with him unto the king.
(II Kings 1:11-15)

These three miracles by Elijah all included fire from heaven, which was most likely lightning. It appears that this was Elijah's trademark miracle. It is little wonder that the false Elijah, the second beast of Revelation 13, will seek to imitate the real prophet by imitation of Elijah's credentials using simulations of the prophet's trademark miracle, the infamous fire from heaven. We find an example of that predicted by the Apostle John.

> *13 And he doeth great wonders, so that he maketh fire come down from heaven on the earth in the sight of men,*
>
> *14 And deceiveth them that dwell on the earth by the means of those miracles which he had power to do in the sight of the beast; (Revelation 13:13-14)*

Will any doubt exist in Jewish minds as to whether this prophet is the anticipated Elijah when he evidently appears to have the ability to call fire down from heaven?

Lying Wonders

Some Biblical students believe the fraudulent Elijah will possess Satanic power to perform the apparently supernatural feat of calling down fire (lightning) from heaven. The Bible calls these miraculous signs of the false prophet "lying wonders," which are untruthful because they turn people's faith toward the dragon's counterfeit Messiah. Whether Satan has the power to produce fire from heaven (or lightning strikes) by supernatural means or not can be debated. It is possible some clandestine form of technology will be utilized by the false prophet to produce mesmerizing effects similar to those which modern day magicians are often capable of creating.

The term "lying wonder" may have a double meaning. First, the wonder is a "lie" because it turns men's hearts away from the real Lord. Second, the apparently supernatural feats may be lies because they may be nothing more than magical tricks utilizing the theatrical tactics of present day magicians who conceal the technological methods used in performance of their "feats of magic."

One can legitimately question whether Satan has the ability to supernaturally produce fire from heaven at the hands of the second beast, the false Elijah, because the Devil was evidently unable to produce fire from heaven for Baal at the contest on Mt. Carmel. When Elijah challenged the Baal worshippers to a contest, he seemed to be rather confident that the winner would be the party which could bring down fire from heaven. Has Satan been developing his supernatural powers at a higher level in our day and age? Someone might suggest that since the contest on Mt. Carmel, the Devil has managed to develop some new skills and abilities to manifest his power in supernatural ways, but that supposition appears to be highly dubious. Man has certainly had the time to develop his own technology over the centuries, but it seems rather likely that Satan is still relatively handicapped for a similar contest in which the winner produces fire from heaven. It seems likely the false Elijah will secretly use human technology to produce his lightning strikes.

In the following passages, we are given some information regarding the final pharaoh and his miracles.

> 8   And then shall that Wicked be revealed, whom the Lord shall consume with the spirit of his mouth, and shall destroy with the brightness of his coming:
>
> 9   Even him, whose coming is after the working of Satan with all power and signs and lying wonders,
>
> 10 And with all deceivableness of unrighteousness in

them that perish; because they received not the love of the truth, that they might be saved.
(II Thessalonians 2:8-10)

The term "lying wonders" literally means "wonders of falsehood." The Greek word can be transliterated as "pseudos," which is recognizable as our word "pseudo." While Jesus does not use the qualifier connoting falsehood in His predictions of the future miracles, one can still question the ability of forces of darkness to simulate the works of divine power by supernatural means of their own.

The actual, if not apparent, supernatural wonder of fire from heaven by the false Elijah will probably spellbind Jewish minds, because while Gentiles seek after wisdom and knowledge, the Jew seeks for a sign (I Cor. 1:22). Before the very eyes of Israel and the world, the false Elijah will seem to manifest his Biblical credentials as the genuine Elijah of Malachi's prophecy by his feats of wonder with fire from heaven. The false Elijah will seem to be the genuine article by perfectly fitting the stereotype Israel may be expecting in the Elijah of Biblical prophecy.

In addition to apparently calling down lightning from heaven, Elijah raised the dead to life, as seen in the incident with the widow's dead son. Here again, the false Elijah may appear to verify his credentials through an apparent impartation of life. Whether the false Elijah will be in some way linked with a possible counterfeit resurrection of the false Messiah is not clearly stated in Scripture. The counterfeit Messiah may have some near death, or an actual death, experience from which he will recover, or from which he will appear to be resurrected. The false prophet may utilize that information in his "ministry," in order to direct worship to the false Messiah. The following New

Covenant Scriptures seem to hint at the Elijah-like power of this false prophet with respect to the impartation of life, if the image of the beast is not some known form of technology to the people of the future.

> *14 And deceiveth them that dwell on the earth by the means of those miracles which he had power to do in the sight of the beast; saying to them that dwell on the earth, that they should make an image to the beast, which had the wound by a sword, and did live.*
>
> *15 And he had power to give life unto the image of the beast, that the image of the beast should both speak, and cause that as many as would not worship the image of the beast should be killed.*
> *(Revelation 13:14-15)*

In these passages we see not only the Elijah-like sign of fire from heaven, but we may see another miracle similar to that performed by Elijah, the giving of life. In I Kings 17:17-24 Elijah raised a widow's dead son to life. In the above passage by the Apostle John, the second beast perhaps can be likened to Elijah in that he is said to give life to the image of the beast, if not to the first beast with the sword wound as well. If the false prophet is not connected to the counterfeit Messiah's apparent "resurrection" (or near death "recovery" from the sword wound), it may be that the masquerading false Elijah will appear to be like the genuine Elijah by seemingly bringing the image of the beast to life.

Through these two Elijah-like copy cat types of miracles, that of bringing down lightning from heaven and of apparently being able to revive the dead or impart life, the false prophet may appear to be a legitimate fulfillment of the

Elijah prophecies from the Jewish book of Malachi to the people of the world.

The Pharaoh's Magicians

The pharaoh of Israel's ancient exodus from Egypt hardened his heart at the miraculous plagues produced by the Lord through the hands of Moses and Aaron. The following passages recount this response.

> *11 Then Pharaoh also called the wise men and the sorcerers: now the magicians of Egypt, they also did in like manner with their enchantments.*
>
> *12 For they cast down every man his rod, and they became serpents: but Aaron's rod swallowed up their rods.*
>
> *13 And he hardened Pharaoh's heart, that he hearkened not unto them; as the LORD had said.*
>
> *22 And the magicians of Egypt did so with their enchantments: and Pharaoh's heart was hardened, neither did he hearken unto them; as the LORD had said. (Exodus 7:11-13, 22)*

These Biblical passages seem to suggest that one of the reasons the ancient pharaoh withstood the clear evidence of Israel's powerful Lord was that his own magicians were able to "clone" some of the miracles which occurred at the hands of Moses and Aaron. The exodus pharaoh may have reasoned that if his own magicians could approximate some of the miraculous plagues that Moses and Aaron had brought upon Egypt, then the possibility existed that, given enough time, pharaoh's magicians would somehow have been able to reproduce all of the signs manifested by the two Israelites, at least on some scale.

It may be that the pharaoh of the exodus was privy to some of the tricks his magicians practiced. Some Bible students prefer to believe the ancient Egyptian pharaoh's magicians performed supernatural miracles which imitated the one's from the Lord. It seems credible to conclude that the pharaoh's magicians were exactly capable of living up to the label they have been categorized with by Scripture, the title of "magician." They may have been ancient counterparts to today's magicians who develop skillful methods of producing some miraculous effects by concealing their actual technological techniques employed in their "performances." The ancient pharaoh may have learned some of the "trade secrets" of some of his magical tricksters, and their cunning slight-of-hand tactics. He may have somehow concluded that Moses and Aaron had stumbled onto more technologically skillful magician's trade secrets somewhere outside of Egypt. The exodus pharaoh may have rationalized away the miraculous power of Israel's Lord by attributing it to some form of cunning human ability that approximated the clever human tactics his own magicians used to deceive their audiences in Egypt. The rod into a serpent was evidently a normal magician's trick. Moses confounded this clever stunt by picking up the serpent by the tail, at the Lord's instructions. If Moses had picked the serpent up by the head, he could have depressed the viper's neck as magicians did to induce a catatonic state which turned the serpent into a "rod." Grabbing the serpent by the tail was a clear indication that Moses was not a trickster.

While some prefer to believe pharaoh's magicians used supernatural powers from Satan, it seems very plausible to conclude the magicians were simply cunning, slight-of-hand tricksters who had developed trade secrets for deceiving their audiences. This conclusion can be seen by the reaction of Pharaoh's magicians to one of the plagues they could not duplicate.

> *18 And the magicians did so with their enchantments to bring forth lice, but they could not: so there were lice upon man, and upon beast.*
>
> *19 Then the magicians said unto Pharaoh, This is the finger of God: and Pharaoh's heart was hardened, and he hearkened not unto them; as the LORD had said. (Exodus 8:18-19)*

The future false Elijah may be a future "miracle worker" of the same school of magic that our own present day magicians are graduates of. He may be a human with cleverly concealed man-made technology which he utilizes for the purpose of mesmerizing crowds, only he may claim his human tricks are actually supernatural wonders and signs. It does seem likely that through demonic possession some of the Lord's wonders can be simulated (as for example the I Cor. 12 gifts of tongues and the word of knowledge). Revelation 16:13-14 indicates that demons are involved in producing signs, and that these demons are associated with the dragon, beast, and false prophet. A combination of human technology and demonic signs may be associated with the "ministry" of the false Elijah.

Conclusion

The coming false Messiah, the future world pharaoh, will have his "miracle" worker just as the ancient pharaoh of Israel's exodus had his own "magic men." The future pharaoh's false miracle worker will be a counterfeit Elijah. The false prophet may deliberately scheme to be recognized as the Biblically predicted Elijah of the Jewish

Scriptures. With most of the world's population, he will dramatically succeed in his plan of deception. Those who choose to become true followers and believers of the Scriptures will see through his charade, because he has already been unmasked and exposed by the divine author of the Bible, and because the fruits of his ministry will be evil, destructive, blasphemous, and in opposition to the Lord.

The next chapter will examine the time of the future world pharaoh's demonic anointing.

Chapter Four

The Demonic Door

What prevents the Biblically predicted future global dictator from being revealed to the world?

The answer to this question involves a number of factors. One element in particular is the possibility that the future world pharaoh may not even be born yet. On the other hand, supposing the predicted world pharaoh may already have been conceived for birth, there does appear to be in the Holy Bible the existence, if you will, of a divine conspiracy to prevent the coming false Messiah and evil world ruler from making a premature debut. This divine conspiracy to prevent the appearance of the anti-Messiah's worldwide control has often been referred to as a "restraining force," or "restraining power." The person who spearheads this strategy to prevent the pseudo-Messiah from appearing before the appropriate time has been referred to in the Scriptures as the "restrainer." Who and what is restraining the anti-Messiah from appearing before the appointed time?

To understand the Biblical teachings about "what" and "who" restrains the evil false Messiah from being revealed, it is best to begin by examining the Scriptures relating to that issue. In addition, it may also be helpful to examine various theories that have existed among students of

Biblical prophecy about the restrainer.

The Thessalonian Briefing

The Thessalonian church received some important teaching from the Pharisee known as the Apostle Paul. That church had been instructed personally by Paul concerning a very important future event, the Second Coming of Israel's Messiah. Involved in that event is a resurrection of deceased, faithful believers, and a simultaneous sudden snatching away of living followers of the Jewish Messiah from earth into the heavens to be with Jesus. The Apostle described that event in detail in his first letter to the Thessalonians.

> *13 But I would not have you to be ignorant, brethren, concerning them which are asleep, that ye sorrow not, even as others which have no hope.*
>
> *14 For if we believe that Jesus died and rose again, even so them also which sleep in Jesus will God bring with him.*
>
> *15 For this we say unto you by the word of the Lord, that we which are alive and remain unto the coming of the Lord shall not prevent them which are asleep.*
>
> *16 For the Lord himself shall descend from heaven with a shout, with the voice of the archangel, and with the trump of God: and the dead in Christ shall rise first:*
>
> *17 Then we which are alive and remain shall be caught up together with them in the clouds, to meet the Lord in the air: and so shall we ever be with the Lord.*

18 Wherefore comfort one another with these words. (I Thessalonians 4:13-18)

In these passages, the Apostle refers to deceased believers as those who "sleep." In verse 16 the Apostle specifically refers to those who sleep as "the dead in Christ." He indicates first that the dead are resurrected, and then living believers are caught up into the air with those resurrected believers to meet the Lord. This event is popularly called the "rapture." It is the sudden snatching or catching up into heaven of living believers which is preceded by the resurrection of dead believers. Both the dead and living followers of the Messiah Jesus are simultaneously taken up into the clouds at this supernatural event. While the sudden snatching away into the heavens of millions of living and dead true followers of Jesus throughout the centuries may sound preposterous to modern secular minds, the disappearances of Enoch and Elijah in some ways foreshadow this rapture event.

The first sudden disappearance we find recorded in the Scriptures is the disappearance of Enoch. Enoch is listed in the genealogies of Genesis as the seventh name in the genealogical list starting from the first man, Adam. Enoch's disappearance seems to present us with a historical parallel to the rapture. Let us examine a description of this event from the Bible.

21 And Enoch lived sixty and five years, and begat Methuselah:

22 And Enoch walked with God after he begat Methuselah three hundred years, and begat sons and daughters:

> *23 And all the days of Enoch were three hundred sixty and five years:*
>
> *24 And Enoch walked with God: and he was not; for God took him. (Genesis 5:21-24)*

Enoch was not. Where did he go? A New Covenant Scripture informs us that Enoch's earthly existence had a very unusual ending.

> *5 By faith Enoch was translated that he should not see death; and was not found, because God had translated him: for before his translation he had this testimony, that he pleased God. (Hebrews 11:5)*

Enoch did not die. Instead, the writer of Hebrews explained that Enoch was translated so that he would not see death. The great Biblical commentator Matthew Henry states the following about this passage in Hebrews concerning Enoch.

> ... 'he was translated, that he should not see death,' nor any part of him be found upon earth; for God took him, soul and body, into heaven, as he will do those of the saints who shall be found alive at his second coming. (Henry, n.d., Heb. 11:5)

In addition to the Enoch incident, another parallel historical example of the rapture can be found in the life of Elijah, who also did not die. We find that Elijah's life ended in his rising up to heaven during a storm.

> *1 And it came to pass, when the LORD would take up Elijah into heaven by a whirlwind, that Elijah went with Elisha from Gilgal.*

> *11 And it came to pass, as they still went on, and talked, that, behold, there appeared a chariot of fire, and horses of fire, and parted them both asunder; and Elijah went up by a whirlwind into heaven.*
> *(II Kings 2:1, 11)*

The King James translation is evidently somewhat less than literal in translating these passages about Elijah. The Pulpit Commentary makes the following explanation of these passages.

> ... literally, 'and Elijah went up in a storm into the heavens.' There is no mention of a "whirlwind;" and "the heavens" are primarily the visible firmament or sky which overhangs the earth. Elijah, like our Lord, rose bodily from the earth into the upper region of the air, and was there lost to sight.
> (Spence & Exell, n.d., II Kings 2:11)

Evidently, Elijah's translation occurred during a storm, and he was not caught up by a wind. His ascension was apparently similar to the ascension of Jesus.

In connection with the future rapture of living believers, the Apostle Paul had instructed the Thessalonians concerning the coming of the false Messiah, or antichrist. This antichrist will evidently gain international control through his globalized world government. The Thessalonians were indebted to the ministry of the Apostle Paul, because it was he who had brought them the gospel. Paul was a Jew from the tribe of Benjamin (Phil. 3:5), and had, on a previous visit, personally instructed the Thessalonians concerning the appearance of the future world pharaoh, the antichrist. The Apostle gave them some details about the restraining force which was preventing the future dictator from coming to power prematurely.

> 5   Remember ye not, that, when I was yet with you, I told you these things?
>
> 6   And now ye know what withholdeth that he might be revealed in his time.
>
> 7   For the mystery of iniquity doth already work: only he who now letteth will let, until he be taken out of the way.
>
> 8   And then shall that Wicked be revealed, whom the Lord shall consume with the spirit of his mouth, and shall destroy with the brightness of his coming:
> (II Thessalonians 2:5-8)

These passages reveal that there is a "what" which restrains the future evil world ruler from being manifested, and there is a "he" who restrains. The word translated "what" in verse six is a definite article in Greek of the neuter gender. The "he" of verse seven is a definite article in Greek of the masculine gender. This switch in genders has been identified as a critical line of evidence by some students of Scripture.

One commentator makes the following observation about the Thessalonians.

> They knew what this restraining power or influence was–knew it from his previous personal teaching, and therefore he does not here repeat the information. We have not the same knowledge, and so must be contented to conjecture his meaning. Because they knew it so well, we know it so imperfectly.
> (Eadie, 1877, p. 274)

It is because the Thessalonians had been so well informed about the restrainer by the Apostle, that Paul states in an ambiguous manner this information about the restrainer in his second letter to the Thessalonians. This ambiguity has resulted in an overwhelming number of theories concerning the identity of the restraining person and the restraining force.

Some students of the prophetic Scriptures have suggested the following theories concerning the identity of the restrainer, and they do not exhaust the conjectures which have been proposed. As a prelude to the list of ideas for the restrainer, it would perhaps be helpful to consider one of the more popular views. The idea currently in vogue states that the restrainer is the Holy Spirit within the church, and once the true church faithful have been raptured and snatched away from earth by the Messiah, the restraining influence of the church disappears from the world. This position particularly identifies the supposed transition between the neuter and masculine Greek genders as evidence for its position. The reasoning is based on the fact that the Holy Spirit in the New Covenant Scriptures is identified by both masculine and neuter pronouns in the Greek language, which is highly unusual, if not a violation of Greek grammar. The problem with this is that the view actually states it is people who are true believers in the church who constitute the restrainer, so that the restrainer is really considered to be the believers who have the Holy Spirit, the Holy Spirit Himself is really not considered to be the restrainer. This view is worth considering, but will be examined later in this chapter.

In addition to the previous view which has gained great popularity among some circles, there are other theories about the identity of the restrainer.

Other views have stated that the restrainer was the Roman Empire; or Nero; or proconsul Vitellius; or human polity, and those who rule that polity. Others have suggested that civil rule constituted the restrainer. Others state the restrainer is evangelical influence, and others that the German Empire restored by Charlemagne constituted the restrainer. All of these ideas have been proposed as the restrainer of Paul's II Thessalonians letter (Eadie, 1877, pp. 334, 345, 362).

The restrainer is defined as the force and person holding back the appearance of the final evil world dictator. There are other theories, but we will not dwell upon them. We will use Scripture to interpret Scripture in an attempt to arrive at a better understanding concerning the identity of the probable restraint and restrainer.

The What and the Who
The II Thessalonian 2:5-8 passage quoted earlier suggests that the taking away of the restraint results in the revealing, or public debut, of the future global pharaoh's identity to the world.

It is in the vision of the Apostle John, while on the island of Patmos, that we learn the beast ascends out of the bottomless pit (Woods, 1994, p. 47). The first beast of Revelation 13 is none other than the future vile world dictator. The future world pharaoh is described as the enemy of the two prophets of Revelation 11 (who bring plagues upon the earth just as Moses and Aaron brought plagues upon Egypt).

> 7   *And when they shall have finished their testimony, the beast that ascendeth out of the bottomless pit shall make war against them, and shall overcome them, and*

kill them. (Revelation 11:7)

In this passage, the Apostle John states that the beast ascends out of the bottomless pit. It is the first beast of Revelation 13 who rules the global government. The second beast of Revelation 13 is the counterfeit Elijah. How do we know that the first beast ascends out of the bottomless pit? We know it because the first beast of Revelation 13 is described as having seven heads and ten horns. In another passage, John the Apostle again describes the first beast. In Revelation 17 the vision on Patmos preserves for us another description of the creature.

> *3 So he carried me away in the spirit into the wilderness: and I saw a woman sit upon a scarlet coloured beast, full of names of blasphemy, having seven heads and ten horns. (Revelation 17:3)*

Later in the same chapter, John's vision reveals who the beast is.

> *8 The beast that thou sawest was, and is not; and shall ascend out of the bottomless pit, and go into perdition: and they that dwell on the earth shall wonder, whose names were not written in the book of life from the foundation of the world, when they behold the beast that was, and is not, and yet is.*
> *(Revelation 17:8)*

In Revelation 17 we see the same seven headed, ten horned beast which previously appeared as the first beast in Revelation 13, and he ascends out of the bottomless pit.

> *1 And I stood upon the sand of the sea, and saw a beast rise up out of the sea, having seven heads and ten*

> *horns, and upon his horns ten crowns, and upon his heads the name of blasphemy. (Revelation 13:1)*

The description of the first beast of Revelation 13 fits the description of the beast of Revelation 17 who is also described as ascending out of the bottomless pit. Some Biblical students have concluded that Judas Iscariot, who betrayed Jesus, becomes the world pharaoh, because he died and will supposedly ascend out of the bottomless pit. This interpretation is unlikely, and should be rejected, even if the Bible describes both the future antichrist and Judas as men of "perdition." It is likely that the beast which rises out of the bottomless pit is the demon which empowers the future world pharaoh, the demon which anoints the false Messiah. It is very likely that the restraint which has been occurring since the days of the Apostle Paul's letter to the Thessalonians is the restraint or confinement of the demon to the bottomless pit who later empowers the anti-Messiah on its release. When the beast demon is no longer bound in the bottomless pit, then the beast spirit is loosed and ascends out of that abyss. It may be at that time that the false Messiah receives his demonic "anointing" from the newly freed beast demon (Woods, 1994, pp. 101).

We can ask, in light of these details: when does the 'beast' spirit or demon ascend out of the bottomless pit?

To answer this question, let us go to the Apostle John's vision where he saw an unusual event occur at the sounding of the fifth of seven trumpets (Woods, 1994, pp. 76-77).

> *1 And the fifth angel sounded, and I saw a star fall from heaven unto the earth: and to him was given the key of the bottomless pit.*

2 And he opened the bottomless pit; and there arose a smoke out of the pit, as the smoke of a great furnace; and the sun and the air were darkened by reason of the smoke of the pit.

3 And there came out of the smoke locusts upon the earth: and unto them was given power, as the scorpions of the earth have power.

4 And it was commanded them that they should not hurt the grass of the earth, neither any green thing, neither any tree; but only those men which have not the seal of God in their foreheads. (Revelation 9:1-4)

In this part of the Apostle John's vision we specifically see a very dramatic event. A star fallen from heaven to earth. It is a fallen star. Stars come to earth by falling (Ladd, 1972, Rev. 9:1). This star is a person, we learn, for it is referred to with a personal masculine pronoun, and is described as being entrusted with a key which evidently opens the bottomless pit. Out of that pit comes a cloud of smoke full of demonic locusts. We know they are not ordinary locusts because they do not eat grass or vegetation as normal locusts do (Dake, 1977, p. 95). These locusts are probably demons. These are not ordinary or normal locusts as can be seen by the fact that they also have a king over them, which locusts do not have (Dake, 1977, p. 95).

7 And the shapes of the locusts were like unto horses prepared unto battle; and on their heads were as it were crowns like gold, and their faces were as the faces of men.

8 And they had hair as the hair of women, and their teeth were as the teeth of lions.

> *9 And they had breastplates, as it were breastplates of iron; and the sound of their wings was as the sound of chariots of many horses running to battle.*
>
> *10 And they had tails like unto scorpions, and there were stings in their tails: and their power was to hurt men five months.*
>
> *11 And they had a king over them, which is the angel of the bottomless pit, whose name in the Hebrew tongue is Abaddon, but in the Greek tongue hath his name Apollyon. (Revelation 9:7-11)*

The fallen star which is given the key to the abyss is probably an angel on the Lord's side, because the key evidently returns to heaven according to Revelation 20.

> *1 And I saw an angel come down from heaven, having the key of the bottomless pit and a great chain in his hand.*
>
> *2 And he laid hold on the dragon, that old serpent, which is the Devil, and Satan, and bound him a thousand years,*
>
> *3 And cast him into the bottomless pit, and shut him up, and set a seal upon him, that he should deceive the nations no more, till the thousand years should be fulfilled: and after that he must be loosed a little season. (Revelation 20:1-3)*

When this passage about the key to the bottomless pit is compared to the Revelation 9 passage about the key to the bottomless pit, the comparison seems to suggest that the Revelation 9:1 star who is given the key is probably an angel, like the one of the Revelation 20 passage, if not the same angel. Some believe the angel who is given the key is an evil angel, but others doubt the key would be given to Satan or one of Satan's cohorts, for Satan will be later confined to the same pit. The key is evidently returned to heaven after the opening of the abyss in Revelation 9, because the key appears again in Revelation 20. The return of the key to heaven makes it appear rather unlikely that an evil angel is given the key, but it is perhaps possible, as some contend. It is the same bottomless pit of Revelation 9 and Revelation 20 from which the beast demon ascends, and in which Satan is later imprisoned for 1,000 years (Dake, 1977, p. 94). It seems very probable that the beast which empowers the anti-Messiah ascends out of the bottomless pit at the fifth trumpet, along with all of the locust spirits seen in Revelation 9.

We know that the false Messiah will be a "desolator" (Daniel 9:27). Out of the abyss comes a spirit whose name in Greek and Hebrew means "destroyer" and "destruction." Could it be that this "destroyer" who is called "Abaddon" and "Apollyon" is the beast demon who anoints the world tyrant who will masquerade as the false Messiah?

We can now see "what" it is that restrains the beast demon: it is a shaft for the abyss that has a lock which restrains him. That lock has a key, and we know angels, maybe only one trusted angel, will open the lock on that pit. It may be the same angel of Revelation 20:1 who in Revelation 9:1 looses the beast demon from the abyss, although these two passages are separate events.

The word for "pit" in Revelation 9:2 is in the neuter gender, in Greek. So now we know the answer to the "what" which restrains the beast demon who later anoints and controls the future world pharaoh. The "what" is the pit. We can now try to answer the "who" that restrains the beast demon.

The star who is given the key to the pit is an intelligent being, as a simple reading of the passage will demonstrate.

> *1   And the fifth angel sounded, and I saw a star fall from heaven unto the earth: and to him was given the key of the bottomless pit.*
>
> *2   And he opened the bottomless pit; and there arose a smoke out of the pit, as the smoke of a great furnace; and the sun and the air were darkened by reason of the smoke of the pit. (Revelation 9:1-2)*

The star who is given the key to the pit has a personal pronoun in the masculine gender, indicating he is a personal being. The "who" which restrains is possibly the angel, if its allegiance is to the Lord. The problem of the masculine and neuter genders respectively composing the "who" and "what" which restrain the beast demon is potentially solved. The "who" and "what" which restrain the beast are separate items which together constitute the restraint preventing the antichrist's appearance on earth. These two restraining entities are the pit and possibly an angel given the key to the pit.

Restrainer Hermeneutics
Hermeneutics is a word referring to the science of Biblical interpretation. Hermeneutics is concerned with properly explaining passages according to the normal rules of language. A "grammatical-literal-historical-hermeneutic" is concerned with interpreting the words in a Biblical passage according to their ordinary meaning in the context of that

Biblical passage. This "grammatical-literal-historical-hermeneutic" is responsible for the recapturing of the futuristic interpretation of prophetic Scriptures in modern times. It is this proper approach to a Biblical interpretation of the Scriptures which has been credited with the reformation's reclamation of justification by faith theology (Lindsey, 1999, p. 105). This scientific method of Biblical interpretation follows the ordinary rules of language in interpreting the Scriptures. This "grammatical-literal-historical-hermeneutic" is the normal method by which most literature is understood. One can define the "grammatical-literal-historical-hermeneutic" as simply taking the rules for understanding ordinary literature and applying them to the Bible.

Those advocating a rapture which has no intervening signs required, and a rapture which can happen at any-moment, are supporters of the "grammatical-literal-historical-hermeneutic". They advocate the interpretation of prophetic passages according to the ordinary rules of language which govern the interpretation of most documents. This is a proper method of interpretation, and it rejects the "spiritualizing" method which arbitrarily assigns allegorical meanings to words. Those who adopt this scientific and sound method of Biblical interpretation are to be commended.

In identifying the restrainer, any-moment rapture teachers equate the Holy Spirit as a reference to the true believers who compose the true church. The any-moment rapture advocates create for themselves a little difficulty by doing this. The Holy Spirit is omnipresent, and so He cannot really be removed from the earth. The neuter and masculine genders which are used to refer to the restrainer can be applied to the Holy Spirit, but where is the evidence that these pronouns can be applied to the believing Messianic community? This view seems to confuse entities with each other as a way of explaining the restrainer verses of II Thessalonians 2. The logic that is used to support this explanation really appears to be strained. Is the restrainer the Holy Spirit, or is the restrainer the living believers? The claim is made that it is the ministry of the Holy Spirit in believers. The whole theory is a house of cards. If you pull out one part, the whole explanation comes tumbling down.

There seems to be a difficulty explaining how the Holy Spirit can become, instead, the ministry of the Holy Spirit in living believers. The whole interpretation is actually based on what is not said. Those supporting the view of an *any-moment rapture requiring no intervening events* suggest first that the Apostle is referring to the Holy Spirit. Suddenly they switch from the Holy Spirit to "the ministry of the Holy Spirit in living believers." So when the living believers are taken out of the way (rapture rescued by the Messiah), the restraint supposedly disappears from earth. The real issue is the question as to whether it is the Holy Spirit or living believers who constitute the restrainer. Since the Holy Spirit cannot leave earth (He is omnipresent and so everywhere present), it must be the living believers who constitute the restrainer. The ministry of the Holy Spirit in believers is a rather abstract view which creates additional problems. Since the Holy Spirit cannot be removed, the ministry of the Holy Spirit would resume in those who are left on earth who come to faith because of the rapture, and through events following the rapture. These new believers would seem to constitute another source for the ministry of the Holy Spirit in believers.

Rather than resorting to the view which states the Holy Spirit is the restrainer, because He is referred to by both masculine and neuter pronouns, and then substituting for the Holy Spirit the explanation that it is really the ministry of the Holy Spirit in living believers which is the restrainer, the solution is to see that two separate entities are involved: 1) a pit and, 2) another entity, possibly an angel, or the One who gives the angel the key. These two separate entities explain the use of the masculine and neuter genders in the explanation of the restrainer. This avoids the highly dubious tactic of switching the Holy Spirit for the "ministry of the Holy Spirit in the body of true living believers." The tactic of explaining the restrainer as the Holy Spirit, but then switching the explanation mid-stream as really some explanation for the ministry of true living believers, appears strained. This has the appearance of rationalizing.

A more natural method of interpretation can be found in explaining the restrainer as a combination of entities, namely: the pit and possibly the angel with the key to the pit, or the One who gives the key to the angel.

To support the idea that it is the Holy Spirit's ministry through present believers which constitutes the restrainer, some pre-tribulationalists go so far as to state that the Lord's dealings with Israel revert to an Old Covenant Dispensation or Economy during Daniel's seventieth week. This seems highly unlikely, in view of the following passage of Scripture.

> *13 In that he saith, A new covenant, he hath made the first old. Now that which decayeth and waxeth old is ready to vanish away. (Hebrews 8:13)*

The hypothesis of a revived former economy has inherent problems. By definition, the older the dispensation or "economy," the more sparse was the quantity of divine self-revelation. Historical revelation which has been given, is history. The death, resurrection and ascension of Jesus cannot be undone (Gundry, 1973, p. 126). Jesus cannot be un-crucified or un-raised from the dead. Perhaps for these reasons the writer of Hebrews predicted the vanishing away of the Old Covenant. To say that an older dispensation or "economy" can be reverted to is like saying the amount of divine revelation can be retracted.

Some claim that future animal sacrifice is evidence of a reversion to an Old Covenant Dispensation or Economy. This does not prove the idea. Part of the reason for that interpretation is probably the denial of the present operation of supernatural gifts of the Spirit in the church today by the man who may have formulated pre-tribulationalism, and possibly by many current heirs to his view. That denial is based on the theory that the supernatural sign gifts of I Corinthians 12 supposedly only operated in early church times in a prolific manner. In order to explain the supernatural sign ministries of the two Revelation Witnesses who smite the earth with plagues as often as they will, it is claimed that an Old Covenant Dispensation or Economy has been reverted to, which conveniently is used to explain the miraculous ministries of the two Witnesses, a miraculous ministry which is denied as being possible now in the church. The fact is that the ministries of the two Witnesses can be fully explained in our own present New Covenant Economy by the supernatural spiritual gifts operative in believers through the ministry of the Holy Spirit today. The sign gifts of the Holy Spirit did not stop in the early church, as will be demonstrated in a future chapter. No switching to an Old Covenant Dispensation or Economy is required. In fact, there does not appear to be in Scripture any precedent or announcement for such a reversion to some prior economy as was predicted for a New Covenant (Jer. 31:31).

The Restrainer

While some have suggested the "fallen star" of Revelation 9:1 is Satan, or a morally fallen angel, it appears more likely to others that the description of "fallen" applies to the concept of physical descent, rather than to a moral or spiritual descent. This seems especially valid since the key to the abyss of Revelation 9 reappears in heaven later, in Revelation 20. Others insist the term "fallen," can nowhere in Scripture be demonstrated as applying to a morally intact, or morally upright angel (Beale, 1999, p. 492). While it is true that the term "fallen," when applied directly to the word "angel" in Greek may always refer to a moral state, the word "fallen" in Revelation 9:1 is applied to a star and so does not indicate a moral state, but may instead indicate a physical location (Charles, 1920, pp. 238-239). The definite article in Greek for "key" would seem to exclude the notion of duplicate keys to the pit. It also seems to some unlikely that the Lord would entrust this vital key to Satan or an evil spirit, seeing that Satan will one day be bound in the same pit for 1,000 years (Revelation 20:1-3), evidently with the same key. It seems unnecessary to believe the lock was changed on the abyss, which could possibly account for a second key. Still, if it is inappropriate to apply the term "fallen" to a morally upright angel, even if it might mean physical rather than moral descent, the restrainer might not be the angel.

It has been suggested that the restrainer could, on the basis of Scripture, be equated with the angel Michael, if the restrainer is indeed an angel. The pit later is secure enough to restrain Satan, so it is powerful enough to restrain the beast spirit. Regardless of the identity of the star which falls in Revelation 9:1, the key to the abyss is given to an angel. Who gives the angel the key? Most likely it is the One who also has other significant keys.

18 I am he that liveth, and was dead; and, behold, I am alive for evermore, Amen; and have the keys of hell and of death. (Revelation 1:18)

7 And to the angel of the church in Philadelphia write; These things saith he that is holy, he that is true, he that hath the key of David, he that openeth, and no man shutteth; and shutteth, and no man openeth; (Revelation 3:7)

*19 And I will give unto thee the keys of the kingdom of heaven: and whatsoever thou shalt bind on earth shall be bound in heaven: and whatsoever thou shalt loose on earth shall be loosed in heaven.
(Matthew 16:19)*

Who has the keys to the kingdom, the key of David, and the keys of hell and of death? It is the Messiah, David's greater Son. The restrainer who restrains the anti-Messiah's anointing beast demon may be the one who opens the abyss, if that angel is not an evil angel. The One who gives the key to the angel who opens the abyss is probably the ultimate or primary restrainer, and is probably Jesus, but even Jesus may transfer the key at the Father's command.

Some contend that the Holy Spirit is the restrainer of II Thessalonians 2, and that He prevents the revealing of the future world pharaoh through His present ministry in true believers, the genuine church. The Scriptural route we have travelled appears to identify the Son of God, the ascended Messiah of Israel, instead as the likely primary restrainer, with even Jesus probably relinquishing the key of the restraining pit temporarily to an angel at the Father's command. This key unlocks the pit out of which the beast demon ascends. It is in the bottomless pit that the beast demon is restrained. When it is released, the beast demon anoints the future false Messiah, the future tyrant of earth.

Conclusion

There is a mystery of godliness, as mentioned by the Apostle Paul in I Timothy 3:16.

> *16 And without controversy great is the mystery of godliness: God was manifest in the flesh, justified in the Spirit, seen of angels, preached unto the Gentiles, believed on in the world, received up into glory. (I Timothy 3:16)*

In contrast to the mystery of godliness, Paul also mentions a mystery of iniquity, or of lawlessness, in II Thessalonians 2:7.

> *7 For the mystery of iniquity doth already work: only he who now letteth will let, until he be taken out of the way. (II Thessalonians 2:7)*

These two mysteries are mysteries in opposing kingdoms, the Kingdom of Heaven, and the Kingdom of Satan.

In Revelation 17 we see the word mystery used again in reference to Babylon. Babylon is Satan's kingdom, and it is the beast spirit who takes control of that ancient kingdom which opposes the Lord through a man.

If the interpretation of the restrainer in this chapter is correct, it would appear that true believers constituting the real church do not constitute the restrainer of II Thessalonians 2. This would mean that the rapture, or snatching away of Messiah's living followers at the rapture, is not required for the revealing and unleashing of the future world dictator. Some people point to the II Thessalonian 2 passage and interpret it as indicating the restrainer must be removed before the false Messiah and world tyrant can appear. Since the restrainer is more probably a pit and an angel (or the pit and the Messiah), this would suggest that the rapture would not be a requirement for the future world pharaoh's revealing.

The evidence seems to suggest that it is not real living followers of the Messiah who restrain the appearance of the antichrist, but a lock on the abyss which is preventing the premature appearance of the last days destroyer who ascends from the bottomless pit.

The spirit named "Abaddon" and "Apollyon" appears to be the king of the demonic locusts. It has been suggested by some that this spirit named "destroyer" and "destruction", according to the Hebrew and Greek names, may also be the same beast demon which anoints the counterfeit Messiah, the predicted future dictator of our planet. He may be the king of the locusts, and the beast does wear crowns.

> *1 And I stood upon the sand of the sea, and saw a beast rise up out of the sea, having seven heads and ten horns, and upon his horns ten crowns, and upon his heads the name of blasphemy. (Revelation (13:1)*

We know the false Messiah's anointing beast demon ascends out of the bottomless pit, and the fifth trumpet is the only opening of the abyss that we observe in the book of Revelation, other than in Revelation 20 when Satan is released at the end of the future 1,000 year millenial kingdom. It appears entirely plausible that it is at the fifth trumpet that the beast demon ascends out of the abyss. If the beast demon's name is Abaddon and Apollyon, the beast's identity is consistent with the meanings of those names. The future pharaoh's anointing demon who ascends out of the abyss will attempt to destroy the people of the Lord, indicating a consistency with the names Abaddon and Apollyon.

It is also interesting that it is not until after the bottomless pit is opened at the fifth trumpet that John begins to identify the beast as one who ascends out of the bottomless pit. We may try to hypothetically suggest another time when the abyss is opened, but one is on dangerous ground when they attempt to add something to the book of Revelation (22:18). The only revealed opening to the abyss in John's apocalypse which fits the time when the beast might ascend out of the bottomless pit is at the fifth trumpet. It is apparently at the opening of the bottomless pit, at the fifth trumpet, that the demon who empowers the future world dictator is finally released, after centuries of being restrained.

The next chapter will examine whether the Apostles exclusively taught the rapture as some claim, or whether Jesus also taught about the rapture.

Chapter Five

Winged Eagles

The Jewish Temple of A.D. 70 was a magnificent structure that was so impressive, even those who disliked Herod admittedly appraised his Temple rebuilding project as the most impressive structure they had ever witnessed on earth. This amazing piece of sprawling architecture formed the backdrop for a conversation between the Rabbi of Nazareth and His followers about A.D. 33.

Surveying the scene from the Mount of Olives, the disciples of Jesus viewed the grandeur of the Herodian Temple as they engaged in conversation with Him. Jesus had astounded them with an almost incomprehensible prediction.

> 2   And Jesus said unto them, See ye not all these things? verily I say unto you, There shall not be left here one stone upon another, that shall not be thrown down. (Matthew 24:2)

To understand the magnitude of this prediction regarding the Temple's total destruction, it might help to understand the amount of labor involved in the rebuilding of the Herodian Jewish Temple. The Apostle John preserves a record of a time Jesus cleansed the Temple at the beginning of Messiah's ministry. As Jesus was standing within the Temple building, just after having overturned the tables of some of the moneychangers and dumping out their money, probably some who were of the Jewish establishment challenged and confronted the fanatically zealous behavior of the Rabbi from Nazareth who had just created such a stir.

18 Then answered the Jews and said unto him, What sign shewest thou unto us, seeing that thou doest these things?

19 Jesus answered and said unto them, Destroy this temple, and in three days I will raise it up.
(John 2:18-19)

It is obvious this statement was totally incredulous to the Rabbi's hearers. The magnificent Temple they were standing in took hundreds, if not thousands, of workers to bring it into its current condition.

20 Then said the Jews, Forty and six years was this temple in building, and wilt thou rear it up in three days?

21 But he spake of the temple of his body.
(John 2:20-21)

Perhaps Rabbi Jesus put off His challengers by deliberately confusing them with His easily misunderstood prediction. He spoke of His own future resurrection, but they thought He spoke of the Herodian Temple they were standing in which had already been in the rebuilding process for forty-six years, and would continue to be rebuilt for approximately three more decades. This amount of labor gives us a small glimpse of the magnitude and grandeur of the

Herodian Temple and its buildings.

In the Olivet Discourse, Jesus declared to His Disciples that none of these stones would remain on top of another. Many of these stones had dimensions greater than those of a human being. The destruction of Israel's national and religious monument, one of the most astounding feats of architecture in the ancient world, immediately had prophetically significant end time implications to the disciples. We hear of those end time concerns in Matthew 24 which records all three parts to the particular questions from the disciples to Jesus. The Olivet Discourse, as it has since been christened, was the Messiah's answer to the disciple's three questions. Historians and students of Scripture have long since pondered the meaning of the predictions made by Jesus that day.

> *3   And as he sat upon the mount of Olives, the disciples came unto him privately, saying, Tell us, when shall these things be? and what shall be the sign of thy coming, and of the end of the world [age]? (Matthew 24:3)*

The English translation here is misleading. The Greek word translated "world" is transliterated as "aionos," which refers to an "age, indefinite time," or "dispensation" (Young, n.d., world). The disciples were not concerned about the end of the earth, they were concerned about the end of the age, and the beginning of the next age with its Messianic Kingdom when Israel would be the central nation, and her Messiah would rule the earth as international global monarch from the throne of David. The destruction of the Temple spoke to them of the end of the Gentile era and the coming golden age for Israel when the Jewish nation would be exalted with her Messiah who would act as world ruler of all rulers.

Most probably, mention of the destruction of Israel's Temple brought to mind Daniel's prophecy concerning the counterfeit Messiah, the future world pharaoh, who would desecrate the Temple of Israel. The disciples were intensely curious about the unfolding of future events which would lead up to and usher in the Messianic Kingdom. In the Olivet Discourse, Jesus gave a series of answers to His disciples which would satisfy the questions they had asked. Among Messiah's answers was a reference to the destruction of the city of Jerusalem, which would result in the destruction of the Temple.

The Olivet Discourse appears in three gospels: Matthew 24, Mark 13, and Luke 21. The Olivet Discourse has been recognized as similar to the Apocalypse, or book of Revelation, because it presents in miniature some of the prophetic, end-time events examined more fully in the book of Daniel and the book of Revelation.

The three gospel writers present different aspects of the reply to the disciples in the Olivet Discourse by Jesus. Each gospel writer concentrates on the items that met his particular goals. The gospel of Matthew appears to be directed primarily to the Jewish reader, and as such, it contains all three aspects of the questions the disciples asked, especially the part concentrating on the close of the age. Since Matthew apparently directs his gospel toward Jewish readers, it was appropriate for him to record the question about signs for the close of this current Gentile era, for those signs would signal the soon start of Israel's Messianic Age, an issue of intense interest to Jewish minds.

The questions the disciples asked covered a rather broad spectrum of events. Included in that list was the issue as to when the Temple would be destroyed. In Luke's account of the Olivet Discourse, we are apparently given a picture of Jerusalem's future destruction, which would be fulfilled in A.D. 70. In Matthew's gospel, more eschatological, or end-time issues, surround the issue of Jerusalem, since Matthew was directed to Jewish readers probably more interested in the Messianic Kingdom.

The Times of the Gentiles
It appears that the Olivet Discourse covers almost the entire Gentile era starting with the first century, which includes the destruction of Jerusalem and its Temple, resulting in the Jewish diaspora (or scattering) into all the nations, down to the end of the Gentile era when Israel would be regathered into its land, occupy Jerusalem again, and see the desecration of the next (to our perspective still future) rebuilt Jewish Temple.

The comprehensive nature of the Olivet Discourse results in some misunderstandings about the meaning of some of the statements Jesus made. None of the gospels alone may fully record the full extent of the details concerning the Gentile era which the Olivet Discourse covers. We know the Olivet Discourse discusses the Gentile era from the time of Jesus and Jerusalem's A.D. 70 destruction down to the signs signaling the beginning of the Messianic Kingdom, because the questions the disciples asked in Matthew 24:3 cover that period of time.

We don't know exactly how the original Olivet Discourse between Jesus and the disciples transpired in its development of answers to each of the disciple's questions. We can't be exactly certain as to how all the details in the discourse were arranged, and to what extent Jesus replied to each of the three questions by the disciples. We can guess and theorize, but we don't have a complete and full account from any one of the gospels alone, and we can only speculate on how the detailed material from each of the three gospels should be harmonized.

Each of the gospels may give somewhat fragmented and disjointed accounts of the Olivet Discourse, resulting in somewhat confused understanding and interpretations concerning what Jesus actually said and meant with particular details, and how each of the three questions were answered by Him. Jesus Himself applied the discourse to the Gentile era, stating that Jerusalem would be trodden down of the Gentiles until that era was consummated.

24 ... and Jerusalem shall be trodden down of the Gentiles, until the times of the Gentiles be fulfilled. (Luke 21:24)

While it has been common in recent years to conclude from the words of Jesus that the times of the Gentiles ended in 1967, with Israel's recapture of the older part of Jerusalem, it should be noted that the city of Jerusalem is still trodden down by Gentiles. Gentiles still control the Temple Mount. Negotiations for a future Palestinian state have included declarations that Jerusalem shall be the capital for the future Palestinian state. Some have suggested that Jerusalem should become internationalized to resolve the feuding over the city as the capital for Israel and as a possible future Palestinian state capital. The Gentile era has not ended yet, but these developments indicate the Gentile era is in its twilight hours. The Gentile era does finally end with the installation of Messiah's kingdom on earth from Israel.

The Gentile era should properly be understood to include the entire seventy weeks recorded in the book of Daniel, which includes the four kingdoms in the image of Nebuchadnezzar's dream (Walvoord, 1990, p. 386). The seventieth week is generally considered by futurists to be the only part of the seventy weeks of Daniel which has been yet unfulfilled. Some students of Scripture feel it is unwarranted to separate Daniel's seventieth week from the other sixty-nine weeks in his book, but there does appear to be internal evidence in the seventy weeks prophecy which warrants such a position. The future seventieth week in Daniel's book is generally considered to be given a detailed prophetic exposition in the book of Revelation by futurists.

Most students of Scripture believe that Jesus included details about events of the seventieth week of Daniel's book in answers to His disciples' three questions during His Olivet Discourse. It can be concluded that Jesus considered the seventieth week of Daniel's book to be a yet future event to Himself. Various Biblical interpreters have arrived at contradictory explanations as to how the details of Daniel's seventieth week are to be fulfilled.

We know from the book of Daniel that the end of the Gentile times initiates the Kingdom rule of the Messiah from Israel. The four kingdoms seen in the image of Nebuchadnezzar's dream cover this entire Gentile era, because during the times of these kingdoms, Israel did not possess autonomous rule. Even today, Israel is pressured by the United States and other powerful nations to follow actions which are not perhaps in their best interests. Other nations seem to be ruling over Israel's affairs even today. Israel does not enjoy a true autonomy with regard to Jerusalem, nor its own covenant land, and is pressured to part with pieces of the land promised to Abraham.

39 And after thee shall arise another kingdom inferior to thee, and another third kingdom of brass, which shall bear rule over all the earth.

40 And the fourth kingdom shall be strong as iron: forasmuch as iron breaketh in pieces and subdueth all things: and as iron that breaketh all these, shall it break in pieces and bruise.

41 And whereas thou sawest the feet and toes, part of potters' clay, and part of iron, the kingdom shall be divided; but there shall be in it of the strength of the iron, forasmuch as thou sawest the iron mixed with miry clay.

42 And as the toes of the feet were part of iron, and part of clay, so the kingdom shall be partly strong, and partly broken.

43 And whereas thou sawest iron mixed with miry clay, they shall mingle themselves with the seed of men: but they shall not cleave one to another, even as iron is not mixed with clay.

44 And in the days of these kings shall the God of heaven set up a kingdom, which shall never be destroyed: and the kingdom shall not be left to other people, but it shall break in pieces and consume all these kingdoms, and it shall stand for ever.

45 Forasmuch as thou sawest that the stone was cut out of the mountain without hands, and that it brake in pieces the iron, the brass, the clay, the silver, and the gold; the great God hath made known to the king what shall come to pass hereafter: and the dream is

*certain, and the interpretation thereof sure.
(Daniel 2:39-45)*

The Gentile era began with the Babylonian capture of Israel under Nebuchadnezzar about the seventh century B.C., and that Gentile era concludes with the earthly physical and bodily return of Jesus the Messiah to rule as the rightful King of planet earth from Israel. This entire Gentile era spans the length of Daniel's seventy weeks, and if there is a gap between the sixty-ninth and seventieth weeks in the book of Daniel, as internal evidence seems to indicate, the Gentile era covers that period of time as well.

The rock which strikes the image in Nebuchadnezzar's dream is to be understood as Messiah's Kingdom which fills all the earth as a political theocracy. While the rock fills the earth in the hearts of true followers of Jesus today, that invisible spiritual Kingdom is to have a future visible manifestation when the Messiah returns to earth and establishes the Messianic Kingdom from Israel. At the time Jesus returns to earth, after conquering the opposition by the armies of the future world pharaoh, Jesus will set up His world government. This world rule of Israel's Messiah is the Kingdom anticipated by Jews throughout history, and was the issue of concern to the disciples when they asked Jesus the three questions found at the start of the Olivet Discourse.

The Historical Scope of the Olivet Discourse
It may be difficult to determine whether the disciples were really asking one question with three parts, or two questions with the second question having two parts, or simply three different questions at the start of the Olivet Discourse. The destruction of the Temple may have immediately brought to mind end of the age events like the com-

ing of the Messiah and the Messianic Kingdom. All three items may have been linked in the minds of the disciples, or it may be that only the last two events were linked in the disciples' minds. However they thought or felt, we can divide the Olivet Discourse into three topics with the signs that Jesus gave assigned to one, two, or even three of the separate items. Jesus answered all three items in the list of questions the disciples offered to Him.

The Olivet Discourse began with the disciples calling attention to the magnificent Temple architecture. In reply to their remarks, Jesus made a specific prediction concerning the destruction of the Temple and its beautiful buildings.

> *1 And as he went out of the temple, one of his disciples saith unto him, Master, see what manner of stones and what buildings are here!*
>
> *2 And Jesus answering said unto him, Seest thou these great buildings? there shall not be left one stone upon another, that shall not be thrown down.*
> *(Mark 13:1-2)*

Questions immediately began to bubble up in the minds of the disciples at such an unusual prediction concerning their magnificent religious monument. The first question the disciples asked in the accounts of the Olivet Discourse by Matthew, Mark, and Luke, was when this predicted destruction would occur.

> *3 And as he sat upon the mount of Olives, the disciples came unto him privately, saying, Tell us, when shall these things be? and what shall be the sign of thy coming, and of the end of the world [age]?*
> *(Matthew 24:3)*

> 3   And as he sat upon the mount of Olives over against the temple, Peter and James and John and Andrew asked him privately,
>
> 4   Tell us, when shall these things be? and what shall be the sign when all these things shall be fulfilled? (Mark 13:3-4)
>
> 7   And they asked him, saying, Master, but when shall these things be? and what sign will there be when these things shall come to pass? (Luke 21:7)

The importance of this first question by the disciples is that it pinpoints a historical incident to which specific answers by Jesus can be assigned. In examining the Olivet Discourse, one must keep in mind that the historical context must include the Jewish Temple's A.D. 70 destruction. This means the Olivet Discourse cannot be interpreted in a manner that ignores the event which sparked the discussion between Jesus and His Disciples: the predicted, thorough destruction, stone by stone, of Israel's religious and national monument by the Roman army under Titus.

While the question about the time of the Temple's destruction by the disciples to Jesus is included in all three accounts of the Olivet Discourse, only Luke appears to give a detailed, direct, prediction of Jerusalem's, and the Temple's, first century destruction which resulted in the diaspora (scattering) of these first century Jews into all nations. Great confusion has arisen among Biblical interpreters on the proper procedure for interpreting the Olivet Discourse, because the time of the Temple's destruction is asked in all three gospels, but only Luke apparently preserves in detail the specific response by Jesus concerning Jerusalem's future demise which results in the Temple's destruction. It appears that only Luke directly and in detail addresses the reply Jesus gave concerning the time of the Herodian Temple's decimation. The result has been for Biblical interpreters to allegorize all statements in the Olivet Discourse dealing with Jerusalem into an answer that deals with the A.D. 70 destruction of the city by Titus. There were two other questions the Disciples asked. Jerusalem is involved in the answers Jesus gave for these two other questions as well.

In a symbolic way, Jerusalem's destruction by Titus in the first century foreshadows the events which occur just prior to the establishment of the Messianic Kingdom. The A.D. 70 destruction of Jerusalem predicts, in a way, the final events which transpire just before Messiah returns to earth. This is an example of history repeating itself. First century events, such as the Romans destroying Israel's Temple, foreshadow events which happen at the end of the Gentile era. Great confusion over these issues has resulted, causing some to say that all or most of the end time predictions contained in the Olivet Discourse were fulfilled completely in the first century when Jerusalem fell to the Roman Armies. This latter view is called "preterism."

Preterists tend to restrict all or most of the Messiah's predictions in the Olivet Discourse to the first century. It has even been suggested that the rapture occurred in the first century. The only way this preterist line of reasoning can be sustained is by spiritualizing or allegorizing away the literality of Biblical prophecy. Israel's resurrection in 1948 suggests that preterists have missed the mark. Israel's 1948 restoration is rationalized away today by some preterists, but was anticipated by futurists who used the grammatical-literal-historical-method of Biblical interpretation. Sir Isaac Newton predicted centuries ago that Israel would be restored as a nation, as other interpreters have also predicted before that restoration even seemed possible (Moore, 1997, p. 20).

The three items in the list of questions the disciples gave to Jesus were: 1) the time of the Temple's destruction, 2) the signs for the Coming of Jesus, 3) the signs for the end of the Gentile era. The second item the disciples asked Jesus to comment about concerned the signs for the return of Jesus. The third item in the disciples' list of questions concerned the time for the ending of the present Gentile age. These two items can be considered two parts to the same question, for in the disciples' minds, the two events were, most likely, closely interrelated. After all, Messiah's return to earth ends the Gentile era, and initiates the Messianic Kingdom which has been anticipated by the Jews for centuries. The following Scriptures predict that future, earthly Messianic Kingdom.

> *5   Behold, the days come, saith the LORD, that I will raise unto David a righteous Branch, and a King shall reign and prosper, and shall execute judgment and justice in the earth.*
>
> *6   In his days Judah shall be saved, and Israel shall dwell safely: and this is his name whereby he shall be called, THE LORD OUR RIGHTEOUSNESS. (Jeremiah 23:5-6)*
>
> *1   Behold, the day of the LORD cometh, and thy spoil shall be divided in the midst of thee.*
>
> *2   For I will gather all nations against Jerusalem to battle; and the city shall be taken, and the houses rifled, and the women ravished; and half of the city shall go forth into captivity, and the residue of the*

people shall not be cut off from the city.

3 Then shall the LORD go forth, and fight against those nations, as when he fought in the day of battle.

4 And his feet shall stand in that day upon the mount of Olives, which is before Jerusalem on the east, and the mount of Olives shall cleave in the midst thereof toward the east and toward the west, and there shall be a very great valley; and half of the mountain shall remove toward the north, and half of it toward the south. (Zechariah 14:1-4)

44 And in the days of these kings shall the God of heaven set up a kingdom, which shall never be destroyed: and the kingdom shall not be left to other people, but it shall break in pieces and consume all these kingdoms, and it shall stand for ever. (Daniel 2:44)

The second item in the list of questions the disciples gave to Jesus establishes another historical period which is distinguishable from the A.D. 70 destruction of the Jewish Temple in the Olivet Discourse. Jesus addressed another of the disciples' concerns when He began to list signs for the end of the Gentile era. The disciples asked Jesus what the signs would be for the end of the current age. The age or era that the disciples were living in, was called by Jesus the "times of the Gentiles," after Jesus predicted the diaspora, or scattering, of Israelites into all the world in the Olivet Discourse.

24 And they shall fall by the edge of the sword, and shall be led away captive into all nations: and Jerusalem shall be trodden down of the Gentiles,

*until the times of the Gentiles be fulfilled.
(Luke 21:24)*

Jesus predicted the Gentile era would be over when Jerusalem was no longer trodden down by Gentiles. That has not yet happened, despite Israel's recapture of the older part of the city of Jerusalem in 1967. We are still presently living in the times of the Gentiles which terminates with the establishment of the physical and visible Messianic Kingdom and the rule of Messiah from the Davidic throne. The Gentile times are not ended as long as Gentiles control the Temple mount.

The scope of the Olivet prophecy by Jesus begins with the first century, continues through history, and lists the signs indicating when the Gentile era will end. There is some overlap by the Olivet Discourse with the seventy weeks of Daniel, and overlap with the times of the four kingdoms in the image of Nebuchadnezzar's dream, also.

As one of the signs for His future coming, Jesus listed the appearance of numerous false Messiahs.

24 For there shall arise false Christs, and false prophets, and shall shew great signs and wonders; insomuch that, if it were possible, they shall deceive the very elect.

25 Behold, I have told you before.

26 Wherefore if they shall say unto you, Behold, he is in the desert; go not forth: behold, he is in the secret chambers; believe it not. (Matthew 24:24-26)

To counteract incorrect ideas which would develop about His own future coming, Jesus then gave analogies which established what a future coming of His would be like.

> *27 For as the lightning cometh out of the east, and shineth even unto the west; so shall also the coming of the Son of man be. (Matthew 24:27)*

Apparently the appearance of Jesus here will be as visible as lightning. The analogy may even suggest what section of the sky Jesus will appear in at this manifestation.

> From the language of this verse probably has been derived the orientation of churches, and the mode adopted of depositing the bodies of deceased Christians, so that they may at the resurrection face the Lord, when he comes from the east.
> (Spence & Exell, n.d., Matt. 24:27)

Then in Matthew Jesus immediately gave another analogy of what His coming would be like.

> *28 For wheresoever the carcase is, there will the eagles be gathered together. (Matthew 24:28)*

Some Biblical interpreters have suggested this analogy of eagles and a corpse is not about the coming of Jesus. Some have intimated instead that it is an analogy about the false Messiahs and their followers. They interpret the "carcass" as the "dead" or false doctrine of a false Messiah. The eagles are said to be vultures or unbelievers who feed or feast on the "dead" teaching or doctrine of a false "Jesus."

Others have suggested this verse about eagles and a carcass depicts the destruction of Jerusalem, and have compared the eagles to the eagles of the Roman standards.

> 'There will be the eagles'] The Roman armies, called so partly from their strength and fierceness, and partly from the 'figure' of these animals which was always wrought on their 'ensigns', or even in brass, placed on top of their ensign-staves. It is remarkable that the Roman fury pursued these wretched men 'wheresoever' they were found. They were a 'dead carcass' doomed to be 'devoured;' and the Roman eagles were the commissioned devourers.
> (Clarke, n.d., Matt. 24:31)

This same passage about the corpse and the eagles is preserved in Luke 17:37. There the context is different, but the analogy there also, by context, appears to apply to the coming of Jesus.

It must be stated that some Scripture passages are, in a way, similar to jokes and puns, because they can sometimes have double (or possibly triple) meanings. Several valid interpretations can sometimes be extracted from a text by exegesis. Sometimes several meanings are read into a text. While the passage about the eagles and the carcass is somewhat ambiguous, it is a text for which many have found significant meaning.

One of the interpretations for Messiah's statement about the eagles and the carcass has found continuity throughout church history. Several early church Fathers found a measure of agreement on the meaning of these passages. Let us examine their statements.

Early Church Fathers Interpret the Eagles and the Carcass

Someone has said that the three key rules for real estate are "location, location, location." It has also been stated the three key principles for interpreting Scripture are "context, context, context." The verse about the eagles and the corpse is found in Matthew, situated just after a sentence which Jesus gave as an analogy to explain His future appearance. Let us examine both verses again and view them together as they appear in the Biblical text.

> *27 For as the lightning cometh out of the east, and shineth even unto the west; so shall also the coming of the Son of man be.*
>
> *28 For wheresoever the carcass is, there will the eagles be gathered together. (Matthew 24:27-28)*

Because the statement about the carcass and eagles is directly next to an analogy about the future appearance of Jesus, it is legitimate to suggest the "eagles and carcass" statement may be another analogy about the appearance of the Messiah. While some have tried to link the eagles and corpse text to verses preceding Matthew 24:27, which are about the false Messiahs, the immediate context for the eagles and corpse verse is the passage in which the appearance of Jesus is compared to lightning. A similar situation is seen in a parallel passage on the eagles and the corpse in Luke 17:37, where the context is apparently about the coming of Jesus. The passage in Luke does not have any nearby texts referring to any false Messiahs. The verse in Luke about the eagles and the corpse is actually given as an answer to the disciples when they ask a question about where someone would be taken. The passage on the eagles and the corpse in Luke even more strongly supports the idea that the verse refers to the coming of Jesus.

St. Thomas Aquinas, an early Christian scholar of great reputation, constructed a commentary from early church

Fathers which preserved their ideas on the gospels. Concerning the eagles and the carcass, here is what he recorded concerning the interpretation by earlier church scholars and leaders on the Matthew 24:28 passage (Aquinas, 2000, Matt. 24:28).

> Another sign He adds of His coming, 'Wheresoever the body is, thither will the eagles be gathered together.' 'The eagles' denote the company of the Angels, Martyrs, and Saints.
> *Chrysostom*

> We may understand by the carcass here, or corpse, which in the Latin is more expressively 'cadaver,' an allusion to the passion of Christ's death.
> *Jerome*

> That we might not be ignorant of the place in which He should come, He adds this, 'Wheresoever the carcase, &c.' He calls the Saints 'eagles,' from the spiritual flight of their bodies, and shews that their gathering shall be to the place of His passion, the Angels guiding them thither; and rightly should we look for His coming in glory there, where He wrought for us eternal glory by the suffering of His bodily humiliation.
> *Hilary*

> They are called eagles whose youth is renewed as the eagle's, and who take to themselves wings that they may come to Christ's passion.
> *Jerome*

In the similar verse found in Luke 17:37, Thomas Aquinas preserves further commentary by earlier Biblical scholars and leaders on that passage (Aquinas, 2000, Luke 17:37).

As if He said, As when a dead body is thrown away, all the birds which feed on human flesh flock to it, so when the Son of man shall come, all the eagles, that is, the saints, shall haste to meet Him.
Cyril

For the souls of the righteous are likened to eagles, because they soar high and forsake the lower parts, and are said to live to a great age. Now concerning the body, we can have no doubt, and above all if we remember that Joseph received the body from Pilate. And do not you see the eagles around the body are the women and Apostles gathered together around our Lord's sepulcher? Do not you see them then, 'when he shall come in the clouds, and every eye shall behold him?' But the body is that of which it was said, 'My flesh is meat indeed;' and around this body are the eagles which fly about on wings of the Spirit, around it also eagles which believe that Christ has come in the flesh.
Ambrose

Amazingly, these early church leaders and scholars suggested an interpretation of the passages about the eagles and the corpse which sounds very much like the rapture. They suggested the corpse refers to the the Passover Lamb of God, Jesus. The eagles are interpreted as believers who feed on that corpse.

Notice what Martin Luther said about that same portion of Scripture (Mueller & Anstadt, 1899, Matt. 24:28).

'Where the carcase is.' That is, where the word is preached and the sacraments are administered, there will also be Christians. Therefore, you need not ask

where the place of Christ's coming will be. I may be where I will, we shall certainly get together. But now it is strange that he compares his kingdom with the carcass of a thief on the gallows. But here Christ is looked upon as nothing but a carcass, or as a condemned, crucified man, and all who believe on him and cling to him, must be regarded as eagles.
Martin Luther

Notice what John Calvin had to say about this text concerning the carcass and eagles (Pringle, n.d., p. 144).

'There also will the eagles be gathered together.'... Christ had manifestly no other design than to call to himself, and to retain in union with him, the children of God, wherever they were scattered... if birds have so great sagacity as to flock in great numbers from distant places to a single 'carcase,' it would be disgraceful in believers not to assemble to the Author of life, from whom alone they derive actual nourishment.
Calvin

The Pulpit Commentary states the following concerning the Matthew passage.

The carcase is Christ, or the body of Christ; the eagles are the saints, or true Christians; these, whatever happens, will, with keen spiritual sight, always be able to discern Christ and his body, and to flock thereto. He calls himself 'ptoma,' because he saves us by his death, and feeds us by his body... Such is the interpretation of many of the Fathers, and it has many analogies in other places of Scripture. Far be it from us to restrict the sphere of Divine prediction, or to assert that any legitimate reference which we may discover was not in the Lord's mind when he spake the words... As a carcase, fall where it may, is immediately observed by the vultures and attracts them, so Christ's coming shall at once be discerned... and draw them into it. (Spence & Exell, n.d., Matt. 24:31)

Could these Scripture passages about the eagle and the corpse refer to the rapture, the sudden snatching away of living believers from the earth to meet the Messiah in the clouds? Could these eagle and corpse passages picture the saints who feed on the Passover Lamb rising to meet Him at His Coming in glory?

Notice the passage states that "wherever" the corpse is the eagles will congregate. Jesus did not state that the corpse was on earth. Is it possible that the Messiah pictured Himself as a slain Passover Lamb (a resurrected 'corpse') in the clouds? For those who would object to such a concept, notice the words of Jesus in John 6.

> *51 I am the living bread which came down from heaven: if any man eat of this bread, he shall live for ever: and the bread that I will give is my flesh, which I will give for the life of the world.*
>
> *52 The Jews therefore strove among themselves, saying, How can this man give us his flesh to eat?*
>
> *53 Then Jesus said unto them, Verily, verily, I say unto you, Except ye eat the flesh of the Son of man, and drink his blood, ye have no life in you.*
>
> *54 Whoso eateth my flesh, and drinketh my blood, hath eternal life; and I will raise him up at the last day.*

55 For my flesh is meat indeed, and my blood is drink indeed.

56 He that eateth my flesh, and drinketh my blood, dwelleth in me, and I in him.

57 As the living Father hath sent me, and I live by the Father: so he that eateth me, even he shall live by me. (John 6:51-57)

The words of Jesus almost sound like cannibalism, but He is depicting Himself as the Passover Lamb which is required for deliverance from the bondage of sin. May I suggest that the Passover Lamb, Jesus the Messiah, is a prerequisite for deliverance from the bondage of the future world pharaoh? It will be those who have partaken of that Lamb, the corpse of Matthew 24:28, who will find the clouds part for them as they soar like eagles to escape the rule of the coming world pharaoh, the antichrist, the false Messiah.

Some may object to these quotations and state that the Olivet Discourse is speaking of the time Messiah returns to earth after the conflict with the world pharaoh. The eagles and corpse metaphor could be claimed to picture the gathering of earth's surviving believers at the time of Messiah's return to this planet. That argument might be convincing to some in Matthew's version of the Olivet Discourse, but it is even there unlikely, and even less convincing in Luke's preservation of the eagles and corpse passage. In addition, there is a specific link of the eagles and corpse passage by an early church Father with the classic I Thessalonians 4:16-17 rapture rescue passage. That quotation is taken from Ambrose, who died about A.D. 397 (Schaff, 1896, p. 192).

... 'the Lord Himself shall descend from heaven at the

voice of the Archangel, and at the trump of God, and they that are dead in Christ shall rise again;' for 'where the body is, there too are the eagles,' where the body of Christ is, there is the truth.
Ambrose

Ambrose links the classic rapture verses of I Thessalonians 4:16-17 specifically with the eagles and corpse metaphor in Messiah's Olivet Discourse.

Conclusion

If the interpretation of the eagles and corpse proposed by church leaders and scholars of past centuries is correct, what we may behold in this passage is the rapture, or sudden snatching away of deceased and living followers of the Messiah. Chrysostom, Ambrose, Calvin, Luther, and others held to this interpretation of the eagles and the corpse as metaphorically picturing the believers and the Messiah. These names are not necessarily bad company to associate with concerning the interpretation of this passage.

The Greek word in Matthew 24:28 for "eagles" is the same Greek word used in Isaiah 40:31 for "eagles" in the Greek translation of the Old Covenant Scriptures, known as the Septuagint.

> *31 But they that wait upon the LORD shall renew their strength; they shall mount up with wings as eagles; they shall run, and not be weary; and they shall walk, and not faint. (Isaiah 40:31)*

If we really wanted to press hard on this passage from Isaiah, we could see the rapture in it. Here in Isaiah 40:31 the Greek Septuagint word for "eagles" is identical to that in Matthew 24:28 where eagles are said to flock to the carcass.

Some will object to this proposed interpretation of the eagles and carcass passages, because it depicts a rapture after the abomination of desolation has occurred. There are other passages of Scripture which suggest a rapture during the seventieth seven of Daniel's seventy weeks, a rapture during the seven years of trouble depicted in the book of Revelation. If the eagles and carcass depict the rapture of I Thessalonians 4:16-17, the hope of a rapture escape of believers before the seven years of turmoil depicted in the book of Revelation evaporates into a mirage of wishful thinking.

While some will object and suggest that an ambiguous and controversial passage such as this about the eagles and carcass is inappropriate for the establishment of a solid Biblical teaching or doctrine, it must be stated that the possibility of the eagles and carcass depicting the Messiah's followers rising to meet Him at the rapture rescue suggests there is danger in committing oneself to endorse a rapture before the seven years of turmoil in the book of Revelation. If the eagles and carcass statement was meant by Jesus to be a parable about the rapture rescue of I Thessalonians 4:16-17, wishful thinking cannot make it otherwise.

If the eagles and carcass are a parabolic statement by Jesus concerning the rapture of I Thessalonians 4, which seems highly likely, a pre-tribulational rapture teaching appears to be an illusory and incorrect hope. I can say this, having at one time myself been someone who accepted and supported a pre-tribulation rapture. It appears from the text of I Thessalonians 4:16-17 (especially in the Greek), that there will be a rapture, but according to Matthew 24:28, not at the time hypothesized by pre-tribulational rapture supporters.

The Greek word used for "eagles" in these passages by

Jesus (Matthew 24:28, Luke 17:37) is identical to the Greek word used in Isaiah 40:31. Some would insist that eagles do not feed on a carcass, so that while the Greek word can be translated "eagles," it would be better translated as some kind of vulture. Jesus did not state that the corpse was on earth. Is it possible that the Messiah pictured Himself as a slain Passover Lamb (a resurrected 'corpse') in the clouds? If this is the final analogy Jesus had in mind, the English translation "eagles" for the Greek word can be considered exceedingly appropriate.

Chapter Six

The Messiah and the Temple

The history of Israel is inextricably linked with the Temple mount and the Temple building that stood there. It is in the same range of mountains where the Temple once stood that Abraham almost slew his son Isaac as a sacrifice on an altar. It is possible Abraham's attempted sacrifice of Isaac happened on the very same mountain where the Temple did stand, as many believe.

When one sees film or video footage about Israel, most often associated in that footage will be pictures of Israelis praying and weeping at the wailing wall. Why the wailing wall? The wailing wall was a retaining wall for the land where the Temple once dominated the horizon. The wailing wall is associated with the Temple, although the wailing wall itself was never a part of the Temple. The wailing wall symbolizes Israel's lost religious monument, the only place where sacrifices were Biblically sanctioned to be offered. Jews congregate to the wailing wall almost in mourning for the loss of the Holy Ancient Temple.

The Temple is the heart of Israel. Israel was established by the Lord originally as a theocracy, not a monarchy. The monarchy came about because Israel wanted a visible, flesh and blood king like other nations. While this displeased Samuel, the Lord nevertheless gave the people what they desired, a king, along with a warning. This situation is documented for us in I Samuel.

4 Then all the elders of Israel gathered themselves together, and came to Samuel unto Ramah,

5 And said unto him, Behold, thou art old, and thy sons walk not in thy ways: now make us a king to judge us like all the nations.

6 But the thing displeased Samuel, when they said, Give us a king to judge us. And Samuel prayed unto the LORD.

7 And the LORD said unto Samuel, Hearken unto the voice of the people in all that they say unto thee: for they have not rejected thee, but they have rejected me, that I should not reign over them.

8 According to all the works which they have done since the day that I brought them up out of Egypt even unto this day, wherewith they have forsaken me, and served other gods, so do they also unto thee.

9 Now therefore hearken unto their voice: howbeit yet protest solemnly unto them, and shew them the manner of the king that shall reign over them.

10 And Samuel told all the words of the LORD unto the people that asked of him a king.

11 And he said, This will be the manner of the king that shall reign over you: He will take your sons, and appoint them for himself, for his chariots, and to be his horsemen; and some shall run before his chariots.

12 And he will appoint him captains over thousands, and captains over fifties; and will set them to ear his ground, and to reap his harvest, and to make his instruments of war, and instruments of his chariots.

13 And he will take your daughters to be confectionaries, and to be cooks, and to be bakers.

14 And he will take your fields, and your vineyards, and your oliveyards, even the best of them, and give them to his servants.

15 And he will take the tenth of your seed, and of your vineyards, and give to his officers, and to his servants.

16 And he will take your menservants, and your maidservants, and your goodliest young men, and your asses, and put them to his work.

17 He will take the tenth of your sheep: and ye shall be his servants.

18 And ye shall cry out in that day because of your king which ye shall have chosen you; and the LORD will not hear you in that day. (I Samuel 8:4-18)

The people asked Samuel for a king, because Samuel's sons didn't follow the ways of the Lord. The people thought the solution to this problem would be to have a king. They wanted to be like the other nations which had kings. The Lord told Samuel to listen to them, but to warn them that taxes and other undesirable consequences would accompany the granting of this request.

The people desired a king anyway.

The first king, Saul, was found deficient in his spiritual walk and was rejected by the Lord as an ancestor for the Messiah. It was King David who came up with the idea for a Temple. David's heart was set on following the God of Israel, and because David honored and reverenced the Lord, David was honored by the Lord in a covenant which established his family as the vehicle through which the Messiah would come.

> *1 And it came to pass, when the king sat in his house, and the LORD had given him rest round about from all his enemies;*
>
> *2 That the king said unto Nathan the prophet, See now, I dwell in an house of cedar, but the ark of God dwelleth within curtains.*
>
> *3 And Nathan said to the king, Go, do all that is in thine heart; for the LORD is with thee.*
>
> *4 And it came to pass that night, that the word of the LORD came unto Nathan, saying,*
>
> *5 Go and tell my servant David, Thus saith the LORD, Shalt thou build me an house for me to dwell in?*
>
> *6 Whereas I have not dwelt in any house since the time that I brought up the children of Israel out of Egypt, even to this day, but have walked in a tent and in a tabernacle.*
>
> *7 In all the places wherein I have walked with all*

the children of Israel spake I a word with any of the tribes of Israel, whom I commanded to feed my people Israel, saying, Why build ye not me an house of cedar?

8   Now therefore so shalt thou say unto my servant David, Thus saith the LORD of hosts, I took thee from the sheepcote, from following the sheep, to be ruler over my people, over Israel:

11   And as since the time that I commanded judges to be over my people Israel, and have caused thee to rest from all thine enemies. Also the LORD telleth thee that he will make thee an house.

12   And when thy days be fulfilled, and thou shalt sleep with thy fathers, I will set up thy seed after thee, which shall proceed out of thy bowels, and I will establish his kingdom.

13   He shall build an house for my name, and I will stablish the throne of his kingdom for ever.

14   I will be his father, and he shall be my son. If he commit iniquity, I will chasten him with the rod of men, and with the stripes of the children of men:

15   But my mercy shall not depart away from him, as I took it from Saul, whom I put away before thee.

16   And thine house and thy kingdom shall be established for ever before thee: thy throne shall be established for ever. (II Samuel 7:1-8, 11-12)

These passages establish the Lord's covenant with David. These passages have Messianic implications, because it is through the Messiah, Jesus, that David's throne is to be established forever.

Matthew and Luke recount the ancestry of the parents of Jesus. Joseph was the legal father of Jesus, but Joseph was not the biological father of Jesus. The lineage of Joseph is probably given in Matthew. That lineage is important, although Joseph was not the biological parent of Jesus. The genealogy in Matthew establishes that Joseph, the legal father of Jesus, was a descendant of King David (Walvoord & Zuck, 1983, p. 18). This establishes the legal right of Jesus to the Davidic throne.

In Luke, the lineage of Mary is probably given through her male parent. Luke probably establishes the fact that Jesus was the biological descendant of King David through Mary.

Matthew's and Luke's genealogies establish that Jesus was legally, and biologically, a descendant of King David. This lineage helps establish the rightful claim of Jesus as Israel's legitimate Messiah. In addition to this lineage showing Jesus as a descendant of King David, there are several hundred Messianic prophecies which were fulfilled by Jesus, many of which would have been impossible for any human to plan a fulfillment for, such as His place of birth in the town of Bethlehem (McDowell, 1999, p. 193).

Mathematically speaking, the odds are virtually impossible for any human being to fulfill the hundreds of Messianic predictions which were fulfilled by Jesus, without some supernatural power behind the events bringing them to pass. Yet despite these supernatural, mathematically impossible fulfillments by Jesus, He has been rejected by the majority of Israel as an illegitimate contender for the position of Messiah.

Despite the hundreds of Messianic prophecies fulfilled by Jesus, the vast majority of Jews even today reject Jesus as their Messiah, despite the astounding evidence to the contrary. It is this amazing situation which has set the stage for what will be one of the greatest hoaxes of world history, Israel's almost wholesale acceptance of a counterfeit Messiah. Israel will accept the person destined to be revealed as the future world pharaoh, as her Messiah, for a temporary period of time, according to New Covenant Biblical prophecy. The rebuilding of Israel's Temple will be one of the major factors which will contribute to this horrendous deception which is predicted to occur in Israel's future.

Who Rebuilds the Temple?

The Temple was King David's brainchild. Before the Temple, the presence of the Lord of Israel and the worship of Him were found within the Tabernacle, a modular, nomadic structure suitable for travel through the wilderness away from Egypt. King David evidently considered it inappropriate for the Creator of the universe to be honored only by a temporary tent, while he as king lived in a permanent building (II Samuel 7:1-12).

The Lord was pleased with David's reverent intentions, but because David was a man of war, the Lord forbid him to build the Temple.

> *7 And David said to Solomon, My son, as for me, it was in my mind to build an house unto the name of the LORD my God:*
>
> *8 But the word of the LORD came to me, saying, Thou hast shed blood abundantly, and hast made great wars: thou shalt not build an house unto my name,*

because thou hast shed much blood upon the earth in my sight.

9 Behold, a son shall be born to thee, who shall be a man of rest; and I will give him rest from all his enemies round about: for his name shall be Solomon, and I will give peace and quietness unto Israel in his days.

10 He shall build an house for my name; and he shall be my son, and I will be his father; and I will establish the throne of his kingdom over Israel for ever. (I Chronicles 22:7-10)

2 Then David the king stood up upon his feet, and said, Hear me, my brethren, and my people: As for me, I had in mine heart to build an house of rest for the ark of the covenant of the LORD, and for the footstool of our God, and had made ready for the building:

3 But God said unto me, Thou shalt not build an house for my name, because thou hast been a man of war, and hast shed blood. (I Chronicles 28:2-3)

2 And Solomon sent to Hiram, saying,

3 Thou knowest how that David my father could not build an house unto the name of the LORD his God for the wars which were about him on every side, until the LORD put them under the soles of his feet.

4 But now the LORD my God hath given me rest on every side, so that there is neither adversary nor evil occurrent.

5 And, behold, I purpose to build an house unto the

> *name of the LORD my God, as the LORD spake unto David my father, saying, Thy son, whom I will set upon thy throne in thy room, he shall build an house unto my name. (I Kings 5:2-5)*

The building of the Temple was to be accomplished by Solomon, but this did not prevent David from making diligent preparations in gathering materials for the future Temple, to make the task of building that great sanctuary easier for his son Solomon.

Perhaps it is because the line of David built the first Temple that there is within Judaism the expectation that the Messiah will build the future Temple. There is undoubtedly within Judaism today the expectation, among some, that the Messiah will rebuild the Temple. In fact, the one who rebuilds the Temple is expected to be the Messiah of Israel, according to a certain segment of observant Jews.

In Zechariah there is a prediction of the Messiah rebuilding the Temple. A common Biblical name for the Messiah is "Branch." The Hebrew is "nezer" from which some believe the word "Nazarene" is derived.

> *12 And speak unto him, saying, Thus speaketh the LORD of hosts, saying, Behold the man whose name is The BRANCH; and he shall grow up out of his place, and he shall build the temple of the LORD:*
>
> *13 Even he shall build the temple of the LORD; and he shall bear the glory, and shall sit and rule upon his throne; and he shall be a priest upon his throne: and the counsel of peace shall be between them both. (Zechariah 6:12-13)*

This Zechariah passage probably predicts the Messiah will build a Millenial Temple. This is most likely the millenial Temple found in Ezekiel 40, or even possibly also the spiritual Temple, the Body of Christ.

> The promise of the future true building of the spiritual temple by Messiah... is an earnest to assure the Jews, that the material temple will be built by Joshua and Zerubbabel, in spite of all seeming obstacles. It also raises their thoughts beyond the material to the spiritual temple, and also to the future glorious temple to be reared in Israel under Messiah's superintendance..." (JFB, n.d., Zech. 6)

Unfortunately, some Israelis today may be applying this Zechariah prophecy about the Messiah building a millenial, or possibly also spiritual, Temple to the current plans for a Temple structure. The current Temple structure in planning may likely be constructed through a treaty mediated with the counterfeit Messiah, the future world pharaoh. Evidence that the false Messiah will be involved in the rebuilding of the tribulational Temple during the seven troublesome years in the book of Revelation can be found in Daniel.

> *27 And he shall confirm the covenant with many for one week: and in the midst of the week he shall cause the sacrifice and the oblation to cease, and for the overspreading of abominations he shall make it desolate, even until the consummation, and that determined shall be poured upon the desolate.*
> *(Daniel 9:27)*

The confirmed covenant is probably an allusion to the rebuilding of the Temple and the reinstitution of animal sacrifice. The length of time for the covenant with many Israelites is stated as one week. It is not a week of days, but of years, and is Daniel's seventieth week. This week of years is most probably the same seven years shown in the vision given to John on Patmos, in the Apocalypse, the book of Revelation. The prince who makes the seven year treaty and who breaks it (Daniel 9:26-27) is the counterfeit Messiah. The treaty he makes evidently concerns the rebuilding of the Temple, possibly causing some observant Jews to identify him as the Messiah when he chooses to reveal himself. Some have speculated that the signing of this treaty will identify who the future world tyrant is, but multiple parties may be involved in the treaty, and the false Messiah may have an ambassador representing himself. For these reasons the treaty may not be a foolproof method for identifying the future world pharaoh. Circumstances may hide the identity of the false prince after the seven year covenant treaty predicted by Daniel is signed.

The Temple and the Temptation
The abomination of desolation is a reference to the above examined prophecy from Daniel 9:27. It refers to a desecration of the Temple which renders that sanctuary useless for Biblically sanctioned Temple practices. Some interpreters believe this passage was fulfilled exclusively by Antiochus Epiphanes (167 B.C.). These interpreters seek to promote the idea that the installation of a pagan altar within the Temple sanctuary by that ruler exhausted the fulfillment of the Daniel prophecy. Usually these interpreters are skeptical of either the divine origin of Scripture, and attempt to rationalize away the prediction as a purported prophecy actually written after the event, or they may be skeptical of the New Covenant Scriptures. The sacrificing of a non-kosher, or unclean, sow on the altar of that pagan shrine by Antiochus Epiphanes certainly does constitute an

apparent fulfillment of the Daniel prophecy, but history does have a way of repeating itself.

Jesus referred to the Daniel 9:27 prophecy in the Olivet Discourse. He referred to it as an event which was still future. While Jesus did not deny the Antiochus incident was in some manner a possible fulfillment of the Daniel prophecy, He did reference the Daniel prophecy as a still future event. This indicates the desecration of the Temple sanctuary by Antiochus may be a foreshadowing of the Daniel prophecy, but that if it is, the Daniel prophecy is not exhausted by the Antiochus incident because Jesus detailed to His disciples the Daniel prophecy as an event He considered to be still future.

15 When ye therefore shall see the abomination of desolation, spoken of by Daniel the prophet, stand in the holy place, (whoso readeth, let him understand:) (Matthew 24:15)

Those who accept the New Covenant Scriptures as divinely inspired have been divided on the proper method for interpreting this text. Some have claimed that Jerusalem's A.D. 70 destruction fulfilled the abomination of desolation prophecy, but further New Covenant evidence clearly demonstrates that the abomination of desolation is an event which has not yet found its ultimate consummation. Like the Antiochus incident, which apparently fulfills the Daniel prophecy, the A.D. 70 destruction of Jerusalem and its Temple also apparently appear to fulfill the same Daniel prophecy.

Some suggest that Israel's rejection of Jesus the Messiah was an abomination which caused the desolation of the Temple in A.D. 70. Jesus did Himself suggest that Israel did not know the time of their own Messiah's visitation, and that the result would be Jerusalem's destruction.

> *41 And when he was come near, he beheld the city, and wept over it,*
>
> *42 Saying, If thou hadst known, even thou, at least in this thy day, the things which belong unto thy peace! but now they are hid from thine eyes.*
>
> *43 For the days shall come upon thee, that thine enemies shall cast a trench about thee, and compass thee round, and keep thee in on every side,*
>
> *44 And shall lay thee even with the ground, and thy children within thee; and they shall not leave in thee one stone upon another; because thou knewest not the time of thy visitation. (Luke 19:41-44)*

While Jerusalem's destruction may in some sense be a shadow of the desolation referred to in Daniel, there is other New Covenant evidence that indicates the abomination of desolation predicted by Daniel and quoted by Jesus refers ultimately in its highest sense to an event which has not yet transpired.

Some students of Biblical prophecy have suggested that the future counterfeit Messiah's false Elijah, who causes the world to worship the image of the beast, sets up an image of the beast in Jerusalem's Temple. The Apocalypse apparently does not indicate specifically whether installation of the beast's image is made inside the Temple. If it is, that could be part of the abomination referred to by Daniel. John describes beast image worship which extends even to the giving of a mark on the hand or forehead, as instigated by the future false Elijah.

> *14 And deceiveth them that dwell on the earth by the means of those miracles which he had power to do in the sight of the beast; saying to them that dwell on the earth, that they should make an image to the beast, which had the wound by a sword, and did live.*
>
> *15 And he had power to give life unto the image of the beast, that the image of the beast should both speak, and cause that as many as would not worship the image of the beast should be killed.*
>
> *16 And he causeth all, both small and great, rich and poor, free and bond, to receive a mark in their right hand, or in their foreheads:*
>
> *17 And that no man might buy or sell, save he that had the mark, or the name of the beast, or the number of his name.*
>
> *18 Here is wisdom. Let him that hath understanding count the number of the beast: for it is the number of a man; and his number is Six hundred threescore and six. (Revelation 13:14-18)*

While this beast image worship is not described as occurring within the Temple, there does seem to be strong evidence it could occur there, despite the fact that it appears to violate the spirit of the ten commandments, specifically the commandment against the worship of graven images.

> *1 And God spake all these words, saying,*
>
> *2 I am the LORD thy God, which have brought thee*

out of the land of Egypt, out of the house of bondage.

3   Thou shalt have no other gods before me.

4   Thou shalt not make unto thee any graven image, or any likeness of any thing that is in heaven above, or that is in the earth beneath, or that is in the water under the earth.

5   Thou shalt not bow down thyself to them, nor serve them: for I the LORD thy God am a jealous God, visiting the iniquity of the fathers upon the children unto the third and fourth generation of them that hate me;

6   And shewing mercy unto thousands of them that love me, and keep my commandments.
(Exodus 20:1-6)

These verses recount the first commandments which compose the decalogue, the ten commandments given to Moses by the Lord. They indicate that the worship of images was forbidden. The false Elijah violates this prohibition in promoting worship of the beast's image.

The Daniel prophecy about the abomination of desolation may be compared to puns or jokes which sometimes have two and possibly three meanings. Here the divine intellect appears to be manifesting a possible triple meaning in the Daniel text. That neither the Antiochus desecration of the Temple, nor the A.D. 70 desolation of the Temple provide the ultimate fulfillments for this abomination prediction in Daniel is demonstrated by the following passages. While the worship of the beast's image in the book of Revelation could be part of the abomination predicted by Daniel, there is another incident which precedes this situation. It is described by the Apostle Paul who stated in his second letter to the Thessalonians that there would be specific events preceding the rapture escape of Messiah's followers. The comments in brackets are my own to clarify an understanding of the text.

> *1   Now we beseech you, brethren, by the coming of our Lord Jesus Christ, and by our gathering together unto him (the rapture rescue of believers),*
>
> *2   That ye be not soon shaken in mind, or be troubled, neither by spirit, nor by word, nor by letter as from us, as that the day of Christ (the day of the rapture rescue) is at hand.*
>
> *3   Let no man deceive you by any means: for that day (the day of the rapture rescue) shall not come, except there come a falling away first, and that man of sin be revealed, the son of perdition;*
>
> *4   Who opposeth and exalteth himself above all that is called God, or that is worshipped; so that he as God sitteth in the temple of God, shewing himself that he is God. (II Thessalonian 2:1-4)*

These Scriptures from Paul the Apostle show strong allusions to the Olivet Discourse of Jesus. It appears that Paul could have borrowed his teaching about that future rapture directly from a transcript of Messiah's list of signs to His disciples. Even some of the language shows similarity to the Olivet Discourse. The Apostle's description of the rapture event and accompanying signs demonstrates that the apostles did not exclusively teach about the rapture, but that Paul's teachings about that event correspond to the teachings of Jesus Himself about the rapture.

Notice how this passage begins. When Paul beseeches the Thessalonians to pay attention, he tells them it is because he is dealing with the same issue about which they have been demonstrating concern: 1) the coming of our Lord Jesus the Messiah, and 2) our gathering together unto Him. These two items are different perspectives or aspects of the same event. First, the Lord comes in the air. Second, the dead and living saints are gathered together unto the Lord who is in the clouds. This is a two-part description of the rapture rescue of believers from earth (I Thessalonians 4:16-17). Paul also seems to indicate in this context that this rapture is part of the "day of the Lord." The King James Version translates it from a textual tradition which transmits the word "Christ." More recent versions of the Scriptures favor a textual tradition which transmits the word "Lord." The "day of Christ" is the "day of the Lord," so there is no real substantive difference in the meaning, other than that "the day of the Lord" has more connotations to the Old Covenant instances of the same phrase.

What Paul seems to be saying is restated in my own words in the following paraphrase and interpretation of the II Thessalonians 2:1-4 Scriptures.

> *1 I beseech you by the rapture event (when Jesus comes in the air and living and dead believers are caught up and gathered unto Him to be with Him),*
>
> *2 that you be not shaken in your composure and calmness concerning that rapture day of rescue because of some letter you have received by someone impersonating us, claiming that the rapture day of rescue has already taken place (that the day of the Lord, or the day of Messiah, has already been started by the snatching away of dead and living believers),*

3 for that day of the Lord which starts with the day of rapture rescue will not begin until there is: 1) first of all a massive apostasy, a widespread falling away from and a departure from the truth of the Biblical faith by people claiming to be followers of the Messiah, and, 2) in addition, the future world tyrant (the future world pharaoh who is the counterfeit Messiah, who is the son of hell and the abyss) must first be unmasked, identified, and exposed by the following act which he will commit;

4 he will set himself up as a being to be worshipped above anything else that is worshipped, including the God of Israel, and he does this ultimately by entering the Jewish sacrificial Temple in Jerusalem where he sets himself down in the Temple as if he were demonstrating his own personal deity and his own divine nature, as though he were actually the Almighty Creator and Lord of Israel and the glorious shekinah cloud of brightness himself.
(Author's Interpretational Paraphrase of
II Thessalonians 2:1-4)

Paul seems to describe in detail the form in which the abomination of desolation predicted by Daniel begins to occur. It is when the counterfeit Messiah, the future world pharaoh, enters the future rebuilt sacrificial Temple in Jerusalem and displays himself as the Almighty Creator. When this happens the abomination of desolation apparently begins. This will be such an unusual and unparalleled event in future world history, that this event itself will immediately provide real followers of the Messiah the unmistakable key to the counterfeit Messiah's actual identity.

Some have supposed that the false Messiah, the antichrist,

will be known to all Biblical followers of Jesus when he mediates or ratifies the treaty or covenant with Israel for the purpose of rebuilding the Jewish Temple on the Temple Mount in Israel. This treaty or covenant will allow the reinstitution of animal sacrifice (Daniel 9:27). The problem with that theory is the potential probability that many nations and individuals may be involved in the signing of that seven year treaty, and the antichrist may have an intermediary who negotiates the treaty. In addition, the possibility exists that the treaty could be made in a secret meeting which ratifies Israel's right to rebuild its Temple and resume animal sacrifice. Certainly the treaty signing event with the parties involved, if it were an item of widespread public knowledge, would likely narrow the field of candidates concerning the precise identity of the counterfeit Messiah, but in II Thessalonians 2:1-4 the Apostle Paul seems to indicate the revealing and unmasking of that future tyrant's identity actually happens when that man of sin assumes the role of the Almighty Creator in the future Jewish sacrificial Temple. The Apostle Paul does not suggest that the pseudo-Messiah's identity will be known by that pretender's role in the signing of the seven year covenant treaty with Israel.

Many separate parties and individuals may be involved in the future signing of the seven year covenant treaty with Israel predicted by Daniel, but only one person will dare to enter the Holy of Holies and display himself as the King of the Universe. There will be no dispute as to the identity of the counterfeit Messiah, the future pharaoh of earth, when he enters the future rebuilt sacrificial Temple in Jerusalem and places himself in the Temple as a being to be worshipped. Unless that event of self-exaltation is discounted by some self-deluded folks living at that time who believe the rapture rescue must precede the antichrist's unmasking, the final identity of the false future pharaoh will be finally known through the counterfeit Messiah's act of self-exaltation inside Israel's Temple.

Two Signs of the Day of the Lord
In II Thessalonians 2:1-4 the Apostle gave two signs for the rapture. First, the apostasy, the departure from the faith will occur. Secondly, the revealing of the beast (the man of sin) must also occur (MacPherson, 1975, p. 109).

Some who teach a sign-less rapture which can happen at any-moment without any required intervening events adopt the pre-tribulation rapture theory. The pre-tribulation rapture theory states that the rapture rescue exodus of Messiah's followers from earth occurs before the seven years of turmoil documented in the book of Revelation. These pre-seven year rapture rescue promoters attempt to define away the concrete signs the Apostle Paul gave for the rapture in II Thessalonians 2:1-4 by defining the "day of the Lord" (or as the King James Version translates it following a different textual tradition, "the day of Christ") as a period of time which excludes the rapture. "The day of the Lord" is defined by some pre-tribulationists as a Biblically defined period of time which follows the rapture rescue event. Some of these same pre-tribulational rapture supporters also define the "day of the Lord" as a period of time which begins when the antichrist's seven-year covenant treaty with Israel is signed which rebuilds the sacrificial Jewish Temple. As we can see, this explanation of the "day of the Lord" defines away the rapture as a part of the "day of the Lord" by its own definition. The passage in II Thessalonians 2:1-4 is then reinterpreted with this customized interpretation which is claimed to be a Biblical definition.

Does the "day of the Lord" begin subsequent to the rapture

exodus? Does the "day of the Lord" exclude the rapture? Does the "day of the Lord" begin with the seven-year treaty between Israel and the antichrist? Is that definition for the "day of the Lord" a Biblical definition as some pre-tribulationists claim? Or is another interpretation used by some pre-tribulationists more correct? Another view states that the "day of the Lord" refers to the event in which Jesus returns to earth at the end of the seven years in the book of Revelation.

With a "day of the Lord" definition which states that the phrase refers to Daniel's seventieth week, a rapture must occur before the seven years detailed in the book of Revelation even start. This issue is so critical, because the entire chronology or time period for the rapture exodus is made to revolve around one's definition of "the day of the Lord." It should be noted that nowhere does the Bible itself explicitly claim that the "day of the Lord" begins with the antichrist's seven-year covenant treaty with Israel. It should also be noted that nowhere does the Bible explicitly claim that the rapture rescue of Messiah's followers from earth is excluded from the "day of the Lord." This definition which excludes the rapture from the "day of the Lord" is an assumption made by pre-tribulationists to support their own position. The "day of the Lord" definition arrived at by those pre-tribulationists is really a matter of subjective interpretation and opinion.

So how do we solve this dilemma for finding a Biblical definition for the "day of the Lord?" Since one's definition for that phrase determines whether one believes the rapture rescue exodus from earth occurs before the seven years of Revelation, or transpires during the seven years of the book of Revelation, finding a Biblically based definition for the "day of the Lord" is of absolutely critical importance. One cannot underestimate the pivotal results which develop in a person's beliefs about the rapture exodus which hinge on one's definition for "the day of the Lord." Consequently, finding a Biblical basis for the phrase "day of the Lord" may determine one's view as to when the rapture actually takes place. Of course, we cannot exclude the possibility of personal bias which might determine the outcome for one's personal definition of the "day of the Lord." If one wishes to believe in a pre-tribulational rapture exodus of Messiah's followers, obviously that person will decide to accept a definition in harmony with their preconceived ideas. All the Biblical evidence in the world will not change the ideas of someone who won't allow the Biblical evidence to shape their position. As the wise proverb states, "A man convinced against his will is of the same opinion still."

Since we are particularly concerned with the phrase "day of the Lord," or "day of Christ" (KJV), as used by the Apostle Paul, let us examine the Apostle Paul's uses of that phrase to determine whether he defines that phrase in a manner which excludes the rapture. Let us begin by reviewing the passages of Scripture where one's definition for the "day of the Lord", or "day of Christ" (KJV), is of such crucial importance in determining whether the rapture has signs.

> *1 Now we beseech you, brethren, by the coming of our Lord Jesus Christ (the coming of Messiah at the rapture rescue? or at the end of the seven years?), and by our gathering together unto him (the rapture rescue of believers),*
>
> *2 That ye be not soon shaken in mind, or be troubled, neither by spirit, nor by word, nor by letter as from us,*

> *as that the day of Christ [day of the LORD] (the day of the rapture rescue? or the seven years of turmoil? or the return to earth of Jesus at the end of the seven years?) is at hand.*
>
> *3 Let no man deceive you by any means: for that day (the day of the rapture rescue? or the seven years of turmoil? or the return to earth of Jesus at the end of the seven years?) shall not come, except there come a falling away first, and that man of sin be revealed, the son of perdition;*
>
> *4 Who opposeth and exalteth himself above all that is called God, or that is worshipped; so that he as God sitteth in the temple of God, shewing himself that he is God. (II Thessalonian 2:1-4)*

The comments in brackets are my own to illustrate the crucial areas where one's definition for the "day of the Lord" may determine how one interprets these passages. If the "day of the Lord" is interpreted to exclude the rapture, then the Apostle is not necessarily talking about signs for the rapture exodus of believers from earth. If the phrase "day of the Lord" includes the rapture, then signs for the rapture do exist. The phrase "day of the Lord" probably preserves the original text, and is used by newer translations. The King James Version is based on a later, and probably corrupted but more interpretational text, which reads "day of Christ." Both phrases mean essentially the same thing, but there are slightly different connotations with the phrase "day of Christ."

Now the question: does the phrase "day of the Lord" (or KJV "day of Christ") include or exclude the rapture rescue? To answer that question, let us examine how the

Apostle Paul uses the phrase elsewhere, because we are particularly concerned with how the Apostle Paul is using that phrase in the above passage which indicates two definite signs for the "day of the Lord." The two signs listed are: 1) the sign of apostasy, and 2) the sign of the revealing of the identity of the future world pharaoh.

Fortunately, we do have examples of the Apostle Paul's use of the phrase "day of the Lord" in other passages of Scripture, and even better yet, the phrase appears in another letter by the Apostle to none other than the Thessalonians! Obviously, the phrase "day of the Lord" in II Thessalonians must contain the identical connotations used by the Apostle previously in I Thessalonians. Let us examine the Apostle's use of that phrase in his first letter.

> *1 But of the times and the seasons, brethren, ye have no need that I write unto you.*
>
> *2 For yourselves know perfectly that the day of the Lord so cometh as a thief in the night.*
> *(II Thessalonians 5:1-2)*

This famous passage of Scripture is perhaps one of the best known rapture texts in the New Covenant Scriptures. The idea of the Lord Jesus coming as a "thief in the night" has long been identified as a rapture text. There are some who may object to this categorization, but a simple reading of the context will reveal that this passage refers to none other event than the rapture. In this passage the "day of the Lord" is compared to "a thief in the night." Let us examine the fuller context to see the fact that this refers to the rapture. The explanatory brackets are mine.

> *14 For if we believe that Jesus died and rose again,*

even so them also which sleep in Jesus will God bring with him.

15 For this we say unto you by the word of the Lord, that we which are alive and remain unto the coming of the Lord (the rapture) shall not prevent them which are asleep.

16 For the Lord himself shall descend from heaven with a shout, with the voice of the archangel, and with the trump of God: and the dead in Christ shall rise first:

17 Then we which are alive and remain shall be caught up together with them in the clouds, to meet the Lord in the air: and so shall we ever be with the Lord (the rapture).

18 Wherefore comfort one another with these words.

1 But of the times and the seasons (of the rapture), brethren, ye have no need that I write unto you.

2 For yourselves know perfectly that the day of the Lord (the rapture) so cometh as a thief in the night.

3 For when they shall say, Peace and safety; then sudden destruction cometh upon them, as travail upon a woman with child; and they shall not escape.

4 But ye, brethren, are not in darkness, that that day (the rapture) should overtake you as a thief.

5 Ye are all the children of light, and the children of the day: we are not of the night, nor of darkness.

6 Therefore let us not sleep, as do others; but let us watch and be sober. (I Thessalonians 4:14-5:6)

The chapter divisions here are not part of the Apostle Paul's original letter, but were artificially introduced centuries later. This is important information which helps us to better understand the continuity in Paul's letter.

The Apostle Paul exhorts the Thessalonians to watch for this "day of the Lord" event. He tells his readers they should not sleep, as do others, but that they should "watch and be sober" so that the "day of the Lord" does not overtake them as a thief. The clear implication is that Paul is speaking about the rapture because it is the Thessalonian believers who are exhorted to watch and not sleep. If the rapture is referred to by the phrase "day of the Lord" in I Thessalonians 5:2, then the rapture is referred to by the phrase "day of the Lord" in II Thessalonians 2:1-4 as well (Van Kampen, 1997, p. 111). If the Apostle Paul is using the phrase "day of the Lord" in I Thessalonians for the rapture, he must be using the same phrase in II Thessalonians for the rapture also. Paul indicates that the term "day of the Lord" includes the rapture event in both his letters to the Thessalonians (I and II Thessalonians). These factors indicate that Paul clearly teaches in II Thessalonians 2:1-4 that two specific signs exist for the rapture: 1) the sign of apostasy, and 2) the sign of the revealing of the identity of the future world pharaoh.

Not only does I Thessalonians 5:1-6 indicate the phrase "day of the Lord" refers to the rapture, the II Thessalonians 2:1-4 passage appears to teach the rapture occurs during the "day of the Lord" from its very own context. From Paul's first letter to the Thessalonians we can see that in Paul's second letter (II Thessalonians 2:1-4) the "gathering" of believers unto the Lord at His coming is "the day of the Lord," it is the "thief in the night" event, it is the same as the rapture mentioned in his first letter to the Thessalonians, and it follows major signs: 1) the sign of apostasy, and 2) the sign of the revealing of the identity of the future world pharaoh.

Where did the Apostle Paul get this information? It is quite possible he got this teaching directly from the Lord, but it is even more likely that Paul had access to a copy of the Olivet Discourse, either as part of one of the gospels, or as a separately circulating document. The evidence for this is quite compelling, both for I Thessalonians 4:14-5:6 and for II Thessalonians 2:1-4.

In Paul's first letter to the Thessalonians (I Thessalonians 4:14-5:6) we see some major parallels to the Olivet Discourse of Matthew 24. Let us compare these passages.

1. Both the Olivet Discourse and I Thessalonians 4:14-5:6 refer to signs.
 • *5:1 But of the times and the seasons, brethren, ye have no need that I write unto you (I Thessalonians).*
 • *24:3 And as he sat upon the mount of Olives, the disciples came unto him privately, saying, Tell us, when shall these things be? and what shall be the sign of thy coming, and of the end of the world (Matthew)?*

2. Both the Olivet Discourse and I Thessalonians 4:14-5:6 use the "thief" analogy.
 • *5:2 For yourselves know perfectly that the day of the Lord so cometh as a thief in the night (I Thessalonians).*
 • *24:43 But know this, that if the goodman of the house had known in what watch the thief would come, he would have watched, and would not have suffered his*

house to be broken up (Matthew).

3. Both the Olivet Discourse and I Thessalonians 4:14-5:6 use the idea of false security.
• *5:3 For when they shall say, Peace and safety; then sudden destruction cometh upon them, as travail upon a woman with child; and they shall not escape (I Thessalonians).*
• *24:48 But and if that evil servant shall say in his heart, My lord delayeth his coming (Matthew).*

4. Both the Olivet Discourse and I Thessalonians 4:14-5:6 refer to the coming of Jesus.
• *4:15 For this we say unto you by the word of the Lord, that we which are alive and remain unto the coming of the Lord shall not prevent them which are asleep (I Thessalonians).*
• *24:42 Watch therefore: for ye know not what hour your Lord doth come (Matthew).*

5. Both the Olivet Discourse and I Thessalonians 4:14-5:6 use the analogy of travail.
• *5:3 For when they shall say, Peace and safety; then sudden destruction cometh upon them, as travail upon a woman with child; and they shall not escape (I Thessalonians).*
• *24:8 All these are the beginning of sorrows [the Greek word translated "sorrows" means labor pangs or travailings] (Matthew).*

6. Both the Olivet Discourse and I Thessalonians 4:14-5:6 use the concept of an unexpected coming for some.
• *5:4 But ye, brethren, are not in darkness, that that day should overtake you as a thief (I Thessalonians).*
• *24:43 But know this, that if the goodman of the house*

had known in what watch the thief would come, he would have watched, and would not have suffered his house to be broken up (Matthew).

7. Both the Olivet Discourse and I Thessalonians 4:14-5:6 refer to a trumpet.
   • *4:16 For the Lord himself shall descend from heaven with a shout, with the voice of the archangel, and with the trump of God: and the dead in Christ shall rise first (I Thessalonians).*
   • *24:31 And he shall send his angels with a great sound of a trumpet, and they shall gather together his elect from the four winds, from one end of heaven to the other (Matthew).*

8. Both the Olivet Discourse and I Thessalonians 4:14-5:6 refer to distraction from watching.
   • *5:6 Therefore let us not sleep, as do others; but let us watch and be sober (I Thessalonians).*
   • *24:42 Watch therefore: for ye know not what hour your Lord doth come (Matthew).*

9. Both the Olivet Discourse and I Thessalonians 4:14-5:6 refer to angels.
   • *4:16 For the Lord himself shall descend from heaven with a shout, with the voice of the archangel, and with the trump of God: and the dead in Christ shall rise first (I Thessalonians).*
   • *24:31 And he shall send his angels with a great sound of a trumpet, and they shall gather together his elect from the four winds, from one end of heaven to the other (Matthew).*

10. Both the Olivet Discourse and I Thessalonians 4:14-5:6 refer to a gathering of Messiah's followers.
    • *4:17 Then we which are alive and remain shall be caught up together with them in the clouds, to meet the Lord in the*

air: and so shall we ever be with the Lord (I Thessalonians).
* *24:31 And he shall send his angels with a great sound of a trumpet, and they shall gather together his elect from the four winds, from one end of heaven to the other (Matthew).*

These ten major parallels between I Thessalonians 4:14-5:6 and the Olivet Discourse in Matthew 24 indicate these Scriptures are very likely referring to the same rapture exodus event.

Paul uses the term "day of the Lord" which has specific Old Covenant Scripture connotations. The term "day of the Lord" occurs frequently in the Jewish Bible with the idea of simultaneous judgment. Some believe it refers to a possibly extended period of time with accompanying judgment (Lindsey, 1999, p. 243). The rapture itself is a judgment. The rapture is the event where Messiah's followers are rescued from this world and rescued from the future devastation which is coming upon the earth. At the rapture unbelievers are judged as unfit to be rescued and are left behind. The leaving behind of unbelievers at the rapture rescue is a judgment of Biblical worldwide proportions with the only previous parallel for the scope of that judgment appearing to be the flood of Noah's day (Matthew 24:37-39).

Paul the Apostle seems to use the phrase "day of the Lord" with connotations derived directly from the Olivet Discourse. This same parallel to the Olivet Discourse is found in Paul's use of the phrase in II Thessalonians 2:1-4 where he gives two specific signs for the "day of the Lord." The two signs are: 1) the sign of apostasy, and 2) the sign of the revealing of the identity of the future world pharaoh. In II Thessalonians 2:1-4 additional parallels can be observed to the Olivet Discourse. The following list examines some of these parallels between Paul's second letter to the Thessalonians and the Olivet Discourse.

1. Both the Olivet Discourse and II Thessalonians 2:1-4 refer to the coming of Jesus.
 • *2:1 Now we beseech you, brethren, by the coming of our Lord Jesus Christ, and by our gathering together unto him (II Thessalonians).*
 • *24:42 Watch therefore: for ye know not what hour your Lord doth come (Matthew).*

2. Both the Olivet Discourse and II Thessalonians 2:1-4 refer to a gathering of Messiah's followers.
 • *2:1 Now we beseech you, brethren, by the coming of our Lord Jesus Christ, and by our gathering together unto him (II Thessalonians).*
 • *24:31 And he shall send his angels with a great sound of a trumpet, and they shall gather together his elect from the four winds, from one end of heaven to the other (Matthew).*

3. Both the Olivet Discourse and II Thessalonians 2:1-4 refer to signs.
 • *2:3 Let no man deceive you by any means: for that day shall not come, except there come a falling away first, and that man of sin be revealed, the son of perdition (II Thessalonians).*
 • *24:3 And as he sat upon the mount of Olives, the disciples came unto him privately, saying, Tell us, when shall these things be? and what shall be the sign of thy coming, and of the end of the world (Matthew)?*

4. Both the Olivet Discourse and II Thessalonians 2:1-4 refer to deception concerning the coming of Jesus.
 • *2:3 Let no man deceive you by any means: for that day shall not come, except there come a falling away first, and*

that man of sin be revealed, the son of perdition (II Thessalonians).
• *24:23 Then if any man shall say unto you, Lo, here is Christ, or there; believe it not.*
24:24 For there shall arise false Christs, and false prophets, and shall shew great signs and wonders; insomuch that, if it were possible, they shall deceive the very elect (Matthew).

5. Both the Olivet Discourse and II Thessalonians 2:1-4 refer to apostasy.
• *2:3 Let no man deceive you by any means: for that day shall not come, except there come a falling away first, and that man of sin be revealed, the son of perdition (II Thessalonians).*
• *24:10 And then shall many be offended [apostasy], and shall betray one another, and shall hate one another (Matthew)(Van Kampen, 1997, p.72).*

6. Both the Olivet Discourse and II Thessalonians 2:1-4 refer to the abomination of desolation as a sign.
• *2:4 Who opposeth and exalteth himself above all that is called God, or that is worshipped; so that he as God sitteth in the temple of God, shewing himself that he is God (II Thessalonians).*
• *24:15 When ye therefore shall see the abomination of desolation, spoken of by Daniel the prophet, stand in the holy place, (whoso readeth, let him understand:) (Matthew).*

This list of parallels between Paul's second letter to the Thessalonians and the Olivet Discourse seem to provide abundant evidence that Paul is indeed referring to that event as the rapture. The Apostle Paul indicates that the rapture is part of the "day of the Lord." He also indicates that the apostasy and the revealing of the man of sin precede the rapture. We must then inquire what these two events are, for they are signs for the rapture exodus.

The Apostasy and the Revealing

What interpretations result from defining the "day of the Lord" in a manner differently than "a period of time beginning with the rapture?" The explanatory brackets indicate the interpretation which is offered by some pre-tribulational supporters.

> *1 Now we beseech you, brethren, by the coming of our Lord Jesus Christ (at the end of the seven years), and by our gathering together unto him (the rapture rescue of believers before the seven years),*
>
> *2 That ye be not soon shaken in mind, or be troubled, neither by spirit, nor by word, nor by letter as from us, as that the day of Christ [day of the LORD] (the day of the seven years of turmoil which begin when the antichrist's seven year covenant treaty is signed) is at hand (has already begun).*
>
> *3 Let no man deceive you by any means: for that day (the day of the seven years of turmoil) shall not come, except there come a falling away first, and that man of sin be revealed, the son of perdition;*
>
> *4 Who opposeth and exalteth himself above all that is called God, or that is worshipped; so that he as God sitteth in the temple of God, shewing himself that he is God. (II Thessalonian 2:1-4)*

What is the resulting interpretation? Here it is in a nutshell: *"Before the seven years of turmoil start, the departure from the faith will occur, and the Man of Sin will be revealed. The Man of Sin is the one who sits in the Jewish Temple displaying himself as the supreme deity."*

Defining the "day of the Lord" as "the seven years of turmoil which begin when the antichrist signs the treaty to rebuild Israel's sacrificial Temple," still results in an interpretation which has the identity of the Man of Sin revealed before the rapture. This interpretation does deny that the identity of the antichrist is revealed by the act of self-exaltation in Israel's future rebuilt sacrificial Temple.

It should be noted that the process of assigning the meaning *seven years of Daniel's seventieth week which begin with the signing of the treaty to rebuild Israel's Temple* to the phrase "day of the Lord," is not a meaning derived from the contexts of Paul's two letters to the Thessalonians. The assignment of this meaning to the phrase "day of the Lord" is really the result of a philosophical evaluation of Daniel's seventieth week and its relation to the coming of Jesus in those Thessalonian passages. This should immediately send up a red flag. Exegesis is an interpretational method which extracts from the text its meaning. On the other hand, eisegesis is a process where a philosophical idea is read into the text to give it a different meaning. It is the latter process, eisegesis, which is being used to define the phrase "day of the Lord" as "Daniel's seventieth week." The contexts of Paul's two letters to the Thessalonians do not endow the phrase with that meaning. On this basis, one should use extreme caution in swallowing the pre-tribulational theory that the phrase "day of the Lord" in Paul's Thessalonian letters refers to "Daniel's seventieth week of seven years," or "tribulation." That definition also contradicts Malachi's prophecy (4:5) that Elijah precedes the "day of the Lord" (Gundry, 1973, pp. 93-95).

Another form of pre-tribulational interpretation is offered below using my explanatory brackets to explain that interpretation. This interpretation suggests that the phrase "day of the Lord" is to be understood to mean *the return of Jesus at the end of the seven years of turmoil in the book of Revelation.*

> *1 Now we beseech you, brethren, by the coming of our Lord Jesus Christ (at the end of the seven years), and by our gathering together unto him (the rapture rescue of believers before the seven years),*
>
> *2 That ye be not soon shaken in mind, or be troubled, neither by spirit, nor by word, nor by letter as from us, as that the day of Christ [day of the LORD] (Messiah's return at the end of the seven years) is at hand (has already begun).*
>
> *3 Let no man deceive you by any means: for that day (Messiah's return at the end of the seven years) shall not come, except there come a falling away first, and that man of sin be revealed, the son of perdition;*
>
> *4 Who opposeth and exalteth himself above all that is called God, or that is worshipped; so that he as God sitteth in the temple of God, shewing himself that he is God. (II Thessalonian 2:1-4)*

This pre-tribulational interpretation results in a denial of any revealing of anything before the rapture. The whole passage is made to sound as if the end of the seven years of turmoil is the issue. Did the Thessalonian believers have the problem of thinking the rapture had already taken place and that the return of Jesus to earth had already taken place also? That seems to be the inference if these passages are interpreted by defining "day of the Lord" as *Messiah's return at the end of the seven years.*

While the interpretation of the phrase "day of the Lord" as *Messiah's return at the end of the seven years* does seem to hold up in this passage, and while it does seem to promote a pre-tribulational rapture idea, and while it does seem to some extent to be derived from the context (first part of verse one), there is a problem. This again is not really exegesis, but another example of eisegesis. The first part of verse one, from the context, appears to be referring not to the return of Jesus at the end of the seven years, but rather to Messiah's coming at the rapture. The coming of Jesus in the air is the first stage of the rapture rescue: 1) Messiah comes in the clouds, 2) Messiah's followers are caught up to be with Him in the air.

The evidence we have previously examined has led us to conclude that the "day of the Lord" is *a period of time which begins with the rapture rescue,* as evidenced by the Apostle Paul's use of that phrase in I Thessalonians 4:14-5:6. Not only does the context indicate in Paul's first letter that he is referring to the rapture, but in Paul's second letter to the Thessalonians the context there as well plainly indicates he is speaking about the rapture. Defining the phrase "day of the Lord" as *a period of time which begins with the rapture rescue,* is not eisegesis, it is an interpretation based on exegesis (extracting from the text and its context the meaning). The phrase "day of the Lord" does have judgment connotations attached to it. Those judgment inferences are satisfied by the rapture. As previously mentioned, the rapture itself is a judgment. The rapture is the event at which the Messiah's followers are rescued from this world and from the future devastation which is coming upon the world. At the rapture unbelievers are judged as being unfit to be rescued, and so they are left behind. The leaving behind of unbelievers at the rapture rescue is a judgment of worldwide proportions. The only previous parallel for the worldwide scope of the rapture judgment appears to be the flood which occurred in Noah's day (Matthew 24:37-39).

There are some who make quite an issue out of the text of II Thessalonians 2:3 in stating that the apostasy and the revealing of the man of sin are basically the same event. The revealing of the man of sin is considered by some to be the ultimate act of apostasy. There may be quite a bit of truth to that, but at least one Greek expert asserts that the apostasy cannot be equated with the revealing of the counterfeit Messiah, on the basis of the word for apostasy.

> The word is found in Acts xxi, 21–a charge against Paul that he taught defection from Moses... This usage shows that by the term spiritual defection is meant... the apostasy precedes, and prepares for the revelation of the Man of Sin. 'The falling away,' therefore, is not the result of the appearance of the Man of Sin, but the antecedent... It is a spiritual falling away, the opposite of that growth in Christian excellence which the apostle commends in them...
> (Eadie, 1877, pp. 265-266)

Some zealous pre-tribulational rapture supporter had attempted to equate the word "apostasy" with the rapture by stating that the "falling away" was a falling away from the planet earth! While the word "apostasy" does indeed contain the idea of departure in it, the inference is that of a spiritual departure away from the truth, not toward the truth as occurs at the rapture. The quote by the Greek commentator above also illustrates the fallacy in the overzealous pre-tribulational interpretation of defining apostasy as the rapture.

Another Greek scholar states concerning the word apostasy in II Thessalonians 2:3, that there is no need to suppose it:

> ... to mean Antichrist himself... nor to regard him as its only cause: rather is he the chief fruit and topstone of the apostasy...
> (Alford, 1856, p. 289)

The apostasy appears to be a response within spiritual groups which possess truth. Members of these groups, rather than embracing that truth, move away and depart from it. The ultimate fruit of this movement away from truth is not only the betrayal of true followers of the Messiah (Matthew 24:10), but the embracing of the counterfeit Messiah at his unveiling.

How will the revealing of the Man of Sin occur? Through the centuries some changes in thinking have occurred about this subject. During the reformation, some interpreted the "temple" of II Thessalonians 2:4 to be the Christian Church. Some concluded the Pope was being spoken of because of the persecution of Protestants by the Catholic Church. Unfortunately, some of the Protestants persecuted other Protestants just as badly when they were able to obtain acceptance in certain geographical areas.

If the "temple" of II Thessalonians 2:4 is interpreted literally, it obviously refers to the Jewish sacrificial Temple in Jerusalem, not to some Christian religious organization or church. After the A.D. 70 destruction of Jerusalem, the "temple" of II Thessalonians 2:4 could only properly infer the future Jewish sacrificial Temple which is predicted elsewhere to be rebuilt by the Scriptures. The obvious allusion is to the Daniel 9:27 prophecy which Jesus referred to in the Olivet Discourse with the phrase "abomination of desolation." One of the earliest interpretations of the passage supposed that the Jewish sacrificial Temple in Jerusalem was being referred to. Ireneaus, an early church Father, believed that was the meaning of the passage.

> Besides he has also pointed out, which in many ways I have shown, that the temple in Jerusalem was made by the direction of the true God... in which temple the adversary shall sit, trying to show himself off as Christ.
> *Ireneaus*
> (Eadie, 1877, pp. 272)

Conclusion

The evidence seems to suggest that Paul was referring to the abomination of desolation when he referred to the Man of Sin as entering the Jewish Temple to display himself as the supreme, divine deity. This coming Temple defiling by a man entering the future rebuilt sacrificial Jewish Sanctuary to display himself as deity appears to be an event which is to transpire before the "day of the Lord." The term "day of the Lord," appears to be used by Paul as a day that begins with the rapture exodus of Messiah's true followers from earth. The revealing of the antichrist's identity appears to be the result of his act of self-exaltation in the future rebuilt Jewish Temple of sacrifice. Many Christian churches and religious organizations will have made a widespread departure from the truth of the Scriptures, perhaps some almost completely abandoning the Bible in some places, maybe not explicitly, but in practice and in theory. This apostasy has been gathering momentum over the years as religious groups claiming to believe the Scriptures have departed farther and farther from the teachings of the Bible.

Has the apostasy taken place? It appears to have begun. It would seem that there will be a much greater departure from the Bible's values and guidelines than we currently observe among "Christian" organizations. The sacrificial Jewish Temple is in the planning stages, but has not, as of this writing, been rebuilt. The two items which Paul seems to have specified as signs for the rapture are not, as of yet, completely fulfilled. It would appear that the two signs Paul designates as prerequisites for the rapture are still, at the very best, only somewhat partially fulfilled.

The next chapter will make a more extensive examination of what Jesus said about His coming. It will examine more evidence to determine if Jesus actually taught about the rapture exodus escape, despite denials by some that He did.

Chapter Seven

The Shofar Blows

> 28 Now learn a parable of the fig tree; When her branch is yet tender, and putteth forth leaves, ye know that summer is near:
>
> 29 So ye in like manner, when ye shall see these things come to pass, know that it is nigh, even at the doors.
>
> 30 Verily I say unto you, that this generation shall not pass, till all these things be done.
> (Mark 13:28-30)

The rabbi from Nazareth uttered these words to His students while He was upon the Mount of Olives, which overlooks the Temple Mount. The Jewish Temple was where animal sacrifices were offered. Jesus had incited His followers into interrogating Him about the end of the age and the time of His coming by an unusual prediction. Jesus had predicted that the magnificent Temple, the rebuilding project of Herod, would so utterly be destroyed that not even one Temple stone would rest upon another.

Rabbi Yeshua, Jesus, gave an extensive reply to His students questions about the time of the Temple's destruction and the end of the age. He listed numerous signs which would: 1) signal the destruction of Jerusalem, 2) the end of the Gentile era, and 3) the time of His return to earth to set up His throne in Jerusalem. Then Jesus uttered the enigmatic words, "... this generation shall not pass, till all these things be done."

The conclusion of a long line of Biblical critics starting with David Strauss is that in this statement, Jesus made an irrecoverable blunder. He made a mistake. He supposedly erred. But did He really?

Numerous and varied approaches have been used to explain these words of Jesus. Skeptics have pounced upon those words to gleefully proclaim that the Messiah, Jesus, was wrong. It has been centuries since Jerusalem's destruction, longer than one generation, but Jesus has not returned (Beasely-Murray, 1993, p. 3).

If one can discredit the predictions of Jesus, then the divine authority they represent can be invalidated. In interpreting the words of Jesus about the generation which would not pass away, critics and skeptics who wish to deny the divine authority Jesus had, attempt to discredit Jesus by interpreting those words in a manner which proves Jesus incorrectly predicted the time of His own coming. While these words of Jesus are not easy to understand concerning the "generation" that will not "pass" away, there are interpretations which indicate the error made by Jesus in that statement is not as easily proven as skeptics suppose we should believe.

A group of interpreters has suggested that the emphasis is to be understood as being placed on the word "generation" in this prophecy by Jesus. Some have suggested that Jesus was indicating the human race would not pass away until everything was fulfilled. Another suggestion is that the word "generation" was to be understood of the Jewish race. Some have taken the word in a moral sense and have interpreted the phrase "generation" to mean "spiritual Israel." This refers to those who believe in Jesus. In that view, "Those who believe in Jesus would not become extinct before all the signs were fulfilled." This view would interpret the statement as an optimistic prediction of hope concerning the historical durability of the Messiah's own followers through the centuries.

Others have interpreted the word "generation" as a reference to unbelievers. This view states that it is the unbelievers who will not pass away until all of what Jesus predicted had been fulfilled.

There are other interpreters who have placed the emphasis on the verb in the sentence. That view has suggested that the generation alive at the time of Jesus would not completely pass away before the signs as a whole began to be fulfilled. John Calvin, a leader of the Reformation, understood the words of Jesus in that sense. Calvin's view is quoted below.

> ... Christ... informs them, that before a single 'generation' shall have been completed, they will learn by experience the truth of what he has said. For within fifty years the city was destroyed and the temple was razed, the whole country was reduced to a hideous desert, and the obstinacy of the world rose up against God... Now the same evils were perpetrated in uninterrupted succession for many ages afterwards, yet what Christ said was true, that, before the close of a single 'generation,' believers would feel in reality, and by undoubted experience, the truth of his prediction; for the apostles endured the same

things which we see in the present day. And yet it was not the design of Christ to promise to his followers that their calamities would be terminated within a short time, (for then he would have contradicted himself, having previously warned them that 'the end was not yet;') but, in order to encourage them to perseverance, he expressly foretold that those things related to their own age... So then, while our Lord heaps upon a single 'generation' every kind of calamities, he does not by any means exempt future ages from the same kind of sufferings, but only enjoins the disciples to be prepared for enduring them all with firmness. (Pringle, 1949, pp. 151-152).

Some would place the emphasis on the word "all" in the sentence, "... this generation shall not pass, till all these things be done." Some would say the word "all" had reference only to Jerusalem's destruction, while others apply the word "all" to include end time events such as Messiah's coming. Preterists, who believe the predictions of Jesus about His own coming were completely fulfilled in the first century, particularly with Jerusalem's A.D. 70 destruction, try to state that this A.D. 70 catastrophe completely fulfilled all of the Messiah's predictions concerning His own coming, with some including the rapture in that event.

The words of Jesus concerning the statement, "... this generation shall not pass, till all these things be done," cannot be used to suggest Jesus was predicting His own return to earth in one generation (as skeptics and as some preterists have tried to apply these words), because Jesus specifically negates the idea that His disciples could know the time of the establishment of the Messianic kingdom when He spoke to His disciples before his ascension into heaven.

> *6 When they therefore were come together, they asked of him, saying, Lord, wilt thou at this time restore again the kingdom to Israel?*
>
> *7 And he said unto them, It is not for you to know the times or the seasons, which the Father hath put in his own power. (Acts 1:6-7)*

This passage in Acts clarifies the fact that Jesus was not in the Olivet Discourse intending to predict His own return to earth in one generation. Skeptics and some preterists have tried to say that Jesus was predicting His own return in one generation, but Jesus clearly demonstrates that is an incorrect interpretation. Jesus knew better what He intended to say in the Olivet Discourse than the skeptics and preterists, and Acts 1:6-7 certainly clarifies His predictions about the Olivet Discourse.

Some have interpreted the "generation" in the prediction by Jesus to be the generation which sees Israel's restoration as a nation. Variations on this concept might suppose "generation" to refer to some generation seeing all of the end time signs, including the abomination of desolation. The generation which sees the abomination of desolation will not pass away until the Gentile era ends and the Millenial Kingdom is set up on earth.

The divine authority of the Olivet Discourse remains intact. Skeptics have not shown an "error" in the prediction made by Jesus, so it is wise to heed the infallible teachings of Jesus concerning the future of the world, and His prediction concerning His own coming and the establishment of the Messianic Kingdom.

Just before Jesus gave the parable of the fig tree to His disciples in the Olivet Discourse, both in Matthew's and Mark's versions of that discussion on the Mount of Olives, the Rabbi from Nazareth mentioned a trumpet.

Jesus and the Trumpet

In ancient Israel a trumpet could refer to a hollowed out horn of an animal which was used as a trumpet, or it could refer to metal silver trumpets. In the case of the hollowed out animal horn, the trumpet was called a "shofar." Jesus mentioned a trumpet at His Coming in Matthew 23:31. Whether that trumpet is a silver trumpet, or a shofar (or some heavenly hybrid material) maybe cannot be known with absolute certainty. Some have tried to turn the trumpet into a metaphor as an analogy for the noise that will occur at the appearance of the Messianic King in the clouds. There seems to be no good reason to reject a literal interpretation of a real trumpet at the appearance in the heavens of the Messianic King. The trumpet appears to be a literal trumpet, whether it be an animal horn, an instrument of metal, or some spiritual material or substance of heaven. At the trumpet sound there is a gathering.

30 And then shall appear the sign of the Son of man in heaven: and then shall all the tribes of the earth mourn, and they shall see the Son of man coming in the clouds of heaven with power and great glory.

31 And he shall send his angels with a great sound of a trumpet, and they shall gather together his elect from the four winds, from one end of heaven to the other. (Matthew 24:30-31)

Gatherings were not unusual when trumpets were blown in Israel (Ex. 19:13, 16, 19; Lev. 23:24, Psalm 81:3-5). The gathering which occurs at the Olivet Discourse trumpet, on the other hand, is extraordinary. That trumpet accompanies a gathering in the clouds!

Did Messiah Mention the Rapture?

In verse thirty of the Matthew 24 Olivet prophecy Jesus mentions an appearance He will make near the end of the Gentile era. He calls this event a "sign." It is interesting that this appearance of Jesus in heaven is referred to as a "sign." It is just another of one of the many signs listed by the Messiah in the Olivet prophecy. Jesus lists numerous signs in this portion of Scripture. Earthquakes, wars, pestilences, famines, persecutions, and false prophets are all "signs." It isn't strange that Jesus would list so many signs, because signs are exactly what His disciples had asked Him for.

The disciples wanted signs for the time the Temple would be destroyed (which took place about forty years later). The disciples also wanted signs for the end of the Gentile era, and they wanted signs for the time Jesus would physically return to earth and visibly set up His expected world monarchy from Jerusalem.

> *3 And as he sat upon the mount of Olives, the disciples came unto him privately, saying, Tell us, when shall these things be? and what shall be the sign of thy coming, and of the end of the world [age]?*
> *(Matthew 24:3)*

Signs were what the disciples wanted, and signs were what the disciples were being given. It is not strange, then, that Jesus would list another "sign" for His disciples.

> *3 And as he sat upon the mount of Olives, the disciples came unto him privately, saying, Tell us, when*

shall these things be? and what shall be the sign of thy coming, and of the end of the world [age]? (Matthew 24:3)

While it's not unusual for Jesus to list signs in the Olivet Discourse, it is unusual for Jesus to list His own appearance in heaven as a sign. Wasn't the appearance of Jesus the event the disciples were inquiring about? Why would Jesus then list the very event the disciples wanted signs for as being a sign itself? According to many interpreters, this particular appearance by Jesus in heaven is the very appearance which is His earthly descent when He will set up His earthly kingdom as King of Earth.

Would Jesus list the very event the disciples were inquiring about as a "sign?" Would Jesus list the very event the disciples were asking signs for as a sign of itself? Or is there some other situation transpiring here that caused Jesus to refer to His appearance as a "sign?" Is it possible this appearance by Jesus in heaven is not the same event about which the disciples are inquiring? Could it possibly be that Jesus is referring to a different visible appearance which He will make in heaven before His coming to set up His earthly kingdom?

What about the rapture? Could this event fit the details of the rapture which is the rescue of Messiah's followers from earth? Could the rapture be listed as a sign of something else? Could the rapture be conceivably listed by Jesus as a "sign" which signals His own future return to earth and the end of the age? Could this event which Jesus describes as His own appearance in heaven be a description of the rapture, and when it occurs, could it be a "sign" that Jesus is soon going to physically return to earth to set up and establish a visible, global, political and spiritual monarchy headquartered in Jerusalem?

There are numerous details about this Matthew 24:30-31 portion of Scripture which seem to suggest that this event is very likely the very same rapture of believers anticipated by students of Biblical prophecy. Some of the most important features about this appearance of Jesus in heaven may be the very things which are not said about it. For example, at this description of an appearance of Jesus in heaven, the Messiah is not pictured as riding a white horse, which He does at the end of the book of Revelation. There is no mention of military hardware, and there is no mention of an army. It is also not stated in the Olivet Discourse that Jesus descends from heaven to touch either the earth or the Mount of Olives as in Zechariah 14:4. Also note that the non-elect, unbelievers, are not stated as being gathered for this appearance of Jesus. Only the elect are stated as being gathered. If the non-elect were being gathered, this might be considered to be the earthly return of Jesus. At Messiah's earthly return there is a gathering of the nations which includes the lost for a judgment of the sheep and goats (Matthew 25:31-46). The elect and the elect only are said to be gathered in the Olivet Discourse of Matthew 24.

> *30 And then shall appear the sign of the Son of man in heaven: and then shall all the tribes of the earth mourn, and they shall see the Son of man coming in the clouds of heaven with power and great glory.*

> *31 And he shall send his angels with a great sound of a trumpet, and they shall gather together his elect from the four winds, from one end of heaven to the other. (Matthew 24:30-31)*

Notice in the above Scriptures that the gathering occurs in the clouds. Notice there is a trumpet. What is fascinating is that these identical features are true of the rapture, the sudden snatching up of dead and living followers of the Messiah to be with Jesus. The details of the rapture are given in Paul's first letter to the Thessalonians.

> *13 But I would not have you to be ignorant, brethren, concerning them which are asleep, that ye sorrow not, even as others which have no hope.*
>
> *14 For if we believe that Jesus died and rose again, even so them also which sleep in Jesus will God bring with him.*
>
> *15 For this we say unto you by the word of the Lord, that we which are alive and remain unto the coming of the Lord shall not prevent them which are asleep.*
>
> *16 For the Lord himself shall descend from heaven with a shout, with the voice of the archangel, and with the trump of God: and the dead in Christ shall rise first:*
>
> *17 Then we which are alive and remain shall be caught up together with them in the clouds, to meet the Lord in the air: and so shall we ever be with the Lord.*
>
> *18 Wherefore comfort one another with these words. (I Thessalonians 4:13-18)*

Notice that the details listed here in the rapture passage of I Thessalonians are identical to the details listed in the gathering which occurs in the Olivet Discourse by Jesus. There is a trumpet. The Lord is not riding a white horse. Messiah is not said to have any military hardware. The Lord does not touch earth or the Mount of Olives. The gathering is exclusively of the elect. The gathering occurs in the clouds. An angel is mentioned as participating in the event.

The specific details of this "gathering" in I Thessalonians 4 are identical to the Biblical details of the gathering of the elect in the Matthew 24:30-31 passage of the Olivet Discourse. Why is this significant? It is highly noteworthy because these details indicate they are two different references to, most likely, the very same event. Despite these identical details, there are some Biblical interpreters who insist these separate references cannot be referring to the same event. Instead, those Biblical interpreters add details to the gathering which occurs in the Olivet Discourse which the Bible does not indicate exist. Those interpreters are trying to transform the Matthew 24:30-31 gathering in the Olivet Discourse into an earthly descent of Jesus, despite the fact that the Biblical text does not state that there is any return to earth. Those interpreters are making their arguments from the Bible's silence. This is significant, because if Jesus does not return to earth at the Matthew 24:30-31 gathering of the elect, what we observe there is a rapture during the seven years of turmoil which are also found in the book of Revelation. Those who teach a rapture before the seven years of turmoil find their pre-tribulational rapture rescue in a little bit of trouble if the Matthew 24 Olivet Discourse gathering of the elect is not an event where the Messiah makes an earthly descent to battle on a white horse or to touch the Mount of Olives as Zechariah describes.

If the Matthew 24:30-31 gathering of the elect in the clouds is the rapture, then that rapture becomes a "sign" that the Messiah's return to earth is to occur some time later. Perhaps that is why Jesus Himself called His Matthew 24:30-31 appearance a "sign." This appearance by Jesus is not the earthly touchdown event about which the disciples inquired, it is a preliminary event different from Messiah's return to earth, so it must be classified as a "sign." It is a "sign," just as earthquakes, pestilences, persecutions, and false Messiahs are signs. The rapture is thus listed by Jesus as one of many "signs" that the end of the Gentile age is about to occur. The rapture is one of many "signs" that the Messiah is soon going to return physically to the planet to set up His Millenial Kingdom and establish His throne.

The Matthew 24:30-31 appearance of Jesus in the clouds is for the rapture, and it is a sign that the physical return of Jesus to earth is about to transpire. When Jesus does return to earth, He sets up His Millenial Kingdom. The return of Jesus to earth does not occur in Matthew 24:30-31. The return there is called a "sign" because it is the rapture. The rapture is a signal that the return to earth of Jesus is not too far off in the future.

The Response on Earth to the Rapture

Jesus, the Rabbi from Nazareth, stated that at His appearance in the heavens, "... then shall all the tribes of the earth mourn, and they shall see the Son of man coming in the clouds of heaven with power and great glory." This statement from Matthew 24:30 indicates that there is a reaction on earth as the Messiah is seen in the clouds. This reaction does not fit the earthly return of Jesus when He comes on a white horse with an army to battle the armies of earth.

19 And I saw the beast, and the kings of the earth, and their armies, gathered together to make war against him that sat on the horse, and against his army. (Revelation 19:19)

At the Olivet Discourse appearance of Jesus there is mourning on earth. In contrast, when Jesus returns to earth at the end of the seven years, there is preparation by earth's forces to war with the rider on the white horse. These different events evoke differing reactions from people on earth.

The mourning on earth at the appearance of Jesus in the clouds was predicted by the prophet Zechariah.

10 And I will pour upon the house of David, and upon the inhabitants of Jerusalem, the spirit of grace and of supplications: and they shall look upon me whom they have pierced, and they shall mourn for him, as one mourneth for his only son, and shall be in bitterness for him, as one that is in bitterness for his firstborn.

11 In that day shall there be a great mourning in Jerusalem, as the mourning of Hadadrimmon in the valley of Megiddon.

12 And the land shall mourn, every family apart; the family of the house of David apart, and their wives apart; the family of the house of Nathan apart, and their wives apart;

13 The family of the house of Levi apart, and their wives apart; the family of Shimei apart, and their wives apart;

14 All the families that remain, every family apart,

and their wives apart. (Zechariah 12:10-14)

Some commentators are agreed that the mourning mentioned here in Zechariah is the same mourning alluded to by Jesus in the Olivet Discourse. The mourning which begins at the rapture appears to possibly increase and crescendo just before Messiah's return to earth. Perhaps it is this mourning which incites the Messiah to mount His white horse and to go forth with His army.

In Matthew 24:30 there is the possibility of translating the word "earth" as "land" instead (Young, n.d., earth). The verse would then read, "... then shall all the tribes of the land mourn, and they shall see the Son of man coming in the clouds of heaven with power and great glory." The phrase "the land" could be construed as a reference for "the promised land." The phrase "the land" recalls to mind the promise of land that the Lord gave to Abraham, specifically the Holy Land, or the "promised land," some of which historically composed Israel under the reigns of David and Solomon. While it is possible to see in this the mourning of Israel's literal tribes, it has been noted that in the Greek Septuagint (a Greek translation of the Old Covenant Jewish Scriptures), the Greek word for "land" never refers to Israelite tribes, but always to earth's nations (Beale, 1999, p. 26). It may be that the Zechariah prophecy is applied by Jesus to include all nations of the earth, as well as being a reference to the tribes of Israel. This could infer that all of earth's nations have been evangelized before Messiah's appearance in the clouds (Matt. 24:14). The mention of these "tribes" evokes thoughts of Israel as well.

> ... all the tribes of the earth mourn (... shall beat their breasts). Not alone the Jews, looking on Him whom they pierced, shall bewail their blindness and impeni-

> tence... but all the nations, the races and peoples who have rejected him whom they ought to have received. (Spence & Exell, n.d., Matt. 24:30)

> ... all the Jewish tribes shall mourn, and many will, in consequence of this manifestation of God, be led to acknowledge Christ and his religion. By... 'of the land,' in the text, is evidently meant here, as in several other places, the land of Judea and its tribes, either its then inhabitants, or the Jewish people wherever found. (Clarke, n.d., Matt. 24:30)

This passage is an amazing prophecy. Jesus seems to allude to Zechariah's prediction about Israel's national mourning. The substance of Zechariah's prediction suggests fulfillment with specific reference to the last days after Israel has been restored to its land. Israel's 1948 resurrection as a geopolitical entity appears to be the threshold for the events which bring about the ultimate fulfillment of this Zechariah prophecy as alluded to by Jesus. At the time of the rapture Israel's tribes see their pierced Messiah in the heavens and mourn for the one they had rejected. This will probably be the time of Israel's national conversion, when they are left behind at that great snatching away of the Messiah's faithful followers from the earth. The rapture may probably be that crisis moment which brings about most of Israel's members to faith in their genuine Messiah, after Israel has been disillusioned by their acceptance of the counterfeit Messiah. The ruling pharaoh of the planet may have his real character finally exposed by the rapture, and this may cause Israelis to turn to Yeshua in repentance and humble contrition.

The inference derived from this combination of descriptions by Jesus in Matthew 24:30-31 has truly mind boggling implications. The apparent description of the rapture, along with the implication to Zechariah's prophecy of Israel's national repentance, leads to the clear inference that the time of the rapture requires Israel to have been re-gathered in their land. An analysis of the Zechariah passage clearly shows that Israel is back in Jerusalem at the very least. This implies that for almost eighteen centuries, the rapture was an impossible event, because Israel from A.D. 70 to 1948 was not a nation. Zechariah clearly infers a Jewish dominance in Jerusalem.

The implications for Israel's restoration to its land as a requirement for the rapture according to Matthew 24:30-31 may be balked at and rejected by many. The fact is that an inference was seen long ago by many Biblical interpreters that Israel had to be re-gathered in their land as a prerequisite for the rapture. This was seen by those Biblical interpreters prior to Israel's 1948 restoration. One Biblical interpreter publicly declared his findings and was branded by an any-moment rapture advocate as being a person who had erred on the Scriptures (Peters, n.d., p. 104). The Biblical interpreter who declared his findings died. The critic who declared him to be in error died. What happened? The rapture did not occur before Israel was re-gathered. The Biblical interpreter declaring Israel had to be re-gathered before the rapture could occur appears to have been correct in his Scriptural analysis. Israel is now re-gathered. One of the inferential signs necessary for the rapture to occur has now transpired. Other signs required for the rapture to occur await fulfillment, such as the rebuilding of Israel's Temple.

While some believe the Olivet verse appears to refer to more than just the twelve tribes of Israel, there is, at the very least, an allusion to the Zechariah 12:12 repentance of

Israel's tribes. Some of the evidence for this is that only Matthew's gospel contains this reference in the Olivet Discourse. Whether correct or in error, according to some Matthew was written in Aramaic originally, and was primarily directed at the Jews. Jerome states the following.

> Matthew published his gospel in Judea, in the Hebrew tongue, for the sake of those of the Jews who believed in Jerusalem.
> *Jerome*
> (Aquinas, 2000, Volume I Part I, p. 8)

First century Israelis who were well versed in the Scriptures would immediately discern the similarity of these words of mourning in Matthew to those in Zechariah 12. At the appearance of the Messiah at the rapture, there appears to be a general repentance and acceptance of Jesus as the legitimate Messiah by Israel's tribes on earth. What hundreds of years of world evangelism could not accomplish within most of Israel happens in moments at Messiah's appearance in the heavens.

The Thessalonian Parallel
The Apostle Paul either had knowledge of the rapture occurring after the abomination of desolation from the Lord directly, or he had obtained that knowledge from a record of the Olivet Discourse itself. If Paul gained his knowledge (or supplemented his knowledge) about the rapture chronology from the Olivet Discourse, it may have been through one of the gospels, or through an independently circulating transcription of the Olivet Discourse. The evidence for that is very convincing, because Paul gave the rapture the same identical chronology which we find given to it by Jesus in the Olivet Discourse. The chronology we find in both Scripture portions (the Olivet

teaching to the Messiah's disciples by Jesus and the Thessalonian teaching by Paul) is that the abomination of desolation precedes the rapture (Matthew 24:15, 30-31; II Thess. 2:1-4).

Paul also refers to the rapture in very similar language to that used by Jesus in His Olivet conversation with His disciples. Both Paul and Jesus call the rapture a "gathering." Paul and Jesus provide a double witness that the abomination of desolation and the revealing of the man of sin precede the rapture. The man of sin is the one who is the counterfeit Messiah, and he is apparently revealed by his act of self-exaltation in the future Jewish sacrificial temple. The unveiling of the identity of the global ruler of earth is apparently accomplished by the abomination of desolation which precedes the rapture.

Jesus and the Rapture?
Those who wish to teach a rapture before the seven years of terrible trial and turmoil in the book of Revelation will deny that the rapture occurs in Matthew 24:30-31, in spite of the evidence to the contrary. While we cannot always convince people of things they don't want to believe, it must be stated that the historical teachings of the early church believers are in agreement with the teaching that in Matthew 24:30-31 we find the rapture, the sudden snatching away of Messiah's followers. As an example, consider the words of John Chrysostom (he died about A.D. 407). His words are quoted concerning the very same Matthew 24:30-31 passage which concerns the heavenly appearance of Jesus (Aquinas, 2000, Matt. 24:30-31).

> That the Lord calls his elect by His Angels pertains to the honour of the elect, and Paul also says that 'they

shall be caught into the clouds;' that is, the angels shall gather together those that have risen, and when they are gathered together, the clouds shall receive them.
Chrysostom

We see in the words of Chrysostom about Matthew 24:30-31 a recognition that this appearance of Jesus is the same event as the appearance of Jesus in I Thessalonians 4:17 when dead and living believers are caught into the clouds to be with the Lord. We can see that an early church father with the stature of Chrysostom interpreted Matthew 24:30-31 as the rapture, the sudden snatching up of true believers (believers living at the time of the rapture along with deceased believers who have been physically resurrected) to meet the Messiah in the clouds.

The meeting of believers with the Messiah occurs in the clouds, in the atmosphere surrounding the earth. It is this event which has been called the "rapture." The word "rapture" is not in the Greek language. The word "rapture" is derived from a word in the Latin translation of the Bible.

Chapter Eight

The Preservation

Passover is a time when Jews recall their deliverance from Pharaoh. The Passover Seder has certain rituals in it which recall the ten plagues and the bitter bondage of Egypt. The ten plagues preceded the exodus, the deliverance of Israel from Egypt. The exodus itself was an escape from Pharaoh's oppressive hardships and brutality which he directed primarily at the Israelites. The ten plagues served the purpose of exposing the powerlessness of the false gods of the Egyptians. It is widely believed that each specific plague had the purpose of demonstrating the superior power of Israel's deity over the supposed power of one or more of the false gods of Egypt.

Pharaoh is an interesting figure in the story of the exodus. Some students of Biblical prophecy see in that ruler of Egypt a picture of the Biblically predicted future evil world ruler who is called the beast and the antichrist. The Bible predicts a future evil ruler who ascends to international power just before the return to earth of Jesus the Messiah. It is during the seven years preceding the much anticipated "golden age of the Messiah," that this future evil world "pharaoh" finally rises to world prominence. This future evil pharaoh's career has astounding similarities to the career of the evil ruler in Egypt who held power at the time of the escape from Egypt by Israel. The similarities are so amazing between the lives of the future wicked pharaoh of earth and the ancient pharaoh of the exodus, that we may in a comparison see simply this: a detailed example of history repeating itself.

As an example of some parallels between the Egyptian pharaoh of the exodus and the future evil world pharaoh, we can consider the plagues. Biblical students have often noted parallels between the plagues on Egypt, and the plagues which descend upon the earth during the rise to world prominence of the future vile world ruler who will masquerade as the Messiah. The plagues in Egypt were directed at the oppressive Egyptians and the hard-boiled pharaoh who apparently lacked any fear or reverence of Israel's Lord. The plagues displayed the awesome power of Israel's Sovereign God, and exposed the powerlessness of Egypt's phony gods.

In the book of Revelation, the last written book of the New Covenant (Jeremiah 33:31-33), the future plagues also appear to be directed at the ungodly followers of the future world pharaoh, while those who follow the Lord are generally preserved from destruction and wrath wrought by the plagues which torment the inhabitants of the earth. This theme of preservation can be noticed in the book of the Apocalypse through the mention of a seal that seems to preserve the elect, and, perhaps, as some speculate, the soon to be elect, from destruction.

> *2   And I saw another angel ascending from the east, having the seal of the living God: and he cried with a loud voice to the four angels, to whom it was given to hurt the earth and the sea,*

3 Saying, Hurt not the earth, neither the sea, nor the trees, till we have sealed the servants of our God in their foreheads.

4 And I heard the number of them which were sealed: and there were sealed an hundred and forty and four thousand of all the tribes of the children of Israel.

5 Of the tribe of Juda were sealed twelve thousand. Of the tribe of Reuben were sealed twelve thousand. Of the tribe of Gad were sealed twelve thousand.

6 Of the tribe of Aser were sealed twelve thousand. Of the tribe of Nephthalim were sealed twelve thousand. Of the tribe of Manasses were sealed twelve thousand.

7 Of the tribe of Simeon were sealed twelve thousand. Of the tribe of Levi were sealed twelve thousand. Of the tribe of Issachar were sealed twelve thousand.

8 Of the tribe of Zabulon were sealed twelve thousand. Of the tribe of Joseph were sealed twelve thousand. Of the tribe of Benjamin were sealed twelve thousand. (Revelation 7:2-8)

4 And it was commanded them that they should not hurt the grass of the earth, neither any green thing, neither any tree; but only those men which have not the seal of God in their foreheads. (Revelation 9:4)

Some Bible interpreters of prophecy suggest that the rapture will have already occurred by the time the false Messiah makes his public debut. These Biblical interpreters follow the doctrine of pre-tribulational teachers, suggesting that real believers will already have risen into the clouds to be with the Lord before the seven dreadful years preceding the millenium begin. If pre-tribulationalism were correct, it would be a wonderful prospect that true followers of the Lord would never have to live through any portion of the most troublesome period in earth's entire history. Unfortunately, as we have already begun to see, there is abundant Biblical evidence to contradict the persecution-free optimism which pre-tribulational, any-moment rapture advocates promulgate. A powerful set of examples contradicting the optimism of pre-tribulational, any-moment rapture rescue advocates can be seen in the prototype example of the Egyptian plagues and Israel's exodus from Egypt. These events appear to be blueprints which illustrate, to some extent, the future seven years and the future exodus of true believers from earth to heaven (the rapture) during the reign of a future world pharaoh.

Much evidence suggests that followers of Israel's Lord must prepare to face devastating plagues which are to come upon the earth. Believers should prepare to pray that they may be accounted worthy to be preserved alive through these coming calamities, just as Israelites were spared and preserved during the period when the ten plagues came upon Egypt.

> *36 Watch ye therefore, and pray always, that ye may be accounted worthy to escape all these things that shall come to pass, and to stand before the Son of man. (Luke 21:36)*

Some believers will be preserved through some of these coming plagues in the same way Israel was preserved through the plagues on Egypt. Any-moment rapture rescue advocates try to put a twist into the instructions Jesus gave and say that believers should pray to escape earth before any of these plagues occur. The prototype for the seven years of turmoil is Israel's time in Egypt under the ten plagues. Like Israel, the true believers are probably going to see devastating plagues ravage the earth before they can hope to escape in the future exodus, the rapture.

The Israelites did not escape Egypt before the ten plagues manifested their terrors. It wasn't until after the ten plagues had been poured out that the Jewish nation escaped from Egypt. In like manner, there is Biblical evidence that the people of the Lord, during the future dreadful seven years of turmoil popularly called the "tribulation," will be preserved from numerous plagues which are poured out upon the earth. Before true believers have their rapture exodus escape from earth to heaven at the appearing in the clouds of the "prophet like unto Moses," the Messiah, many believers will pass through supernatural plagues unscathed like ancient Israel.

We have already examined some specific parallels between the reign of the future evil world pharaoh and the ancient Egyptian pharaoh of the exodus. A comparison of those two careers may enable us to get a better grasp on the events that are going to transpire upon our planet, probably in the not too distant future.

The Future Exodus from Earth
Many Biblical students are anticipating the even at which dead and living true followers of the Messiah ascend into the heavens to be with their Messiah (I Thess. 4:16-17). This event, the rapture rescue, has sometimes been likened to the exodus of Israel from Egypt. The comparison is even more valid if supernatural plagues precede the escape of believers from the earth.

The word "exodus" is Greek, and means the "out" (ex) "road or way" (hodos). The departure of Jesus from earth at His ascension into heaven is also a sort of rapture. The ascension of Jesus into heaven is an event which symbolizes or foreshadows the supernatural rapture and resurrection of the Lord's true followers, so we may properly liken it to an exodus.

In the previous chapter we noted a prominent early church leader who interpreted the Matthew 24:30-31 appearance of the Messiah in the clouds as the rapture of I Thessalonians 4:16-17. Elsewhere we have also noted other evidence that indicates the rapture occurs in the passage which is commonly called the "Olivet Discourse." Prior to that rapture in the Olivet Discourse, Jesus listed a series of plagues and devastating earthly events which will occur. These events have interesting parallels to the plagues which preceded Israel's exodus from Egypt. Jesus predicted that events would precede the rapture. Let us examine some of the predicted events which are shown in the book of Revelation and which correspond to the predictions of Jesus.

In examining the Olivet Discourse which is found in the three synoptic gospels (Matthew, Mark, Luke), we have noted several items which indicate the rapture is probably described by Jesus in that discourse, contrary to the claims by some that Jesus never mentioned the rapture in the Olivet Discourse. Some pre-tribulational rapture rescue advocates even go so far as to say that Jesus was silent on the issue of the rapture. It would appear that these deniers of Jesus as a rapture teacher are incorrect, if the eagles and the corpse is a description of the rapture event. If the eagles and corpse analogy illustrates the moment Messiah's followers rise to meet Him in the clouds, as several ancient church leaders seem to suggest, Jesus did teach about the rapture.

The "sign" of the Son of Man (Matt. 24:30-31) is evidently the rapture. The rapture itself is a sign of Messiah's future return to earth at the end of the present Gentile age. The rapture is also a sign that the future Millenial Kingdom is about to be set up. Jesus appears to have referred to the rapture by the phrase "the sign of the Son of Man."

It should be noted that at the appearance of Jesus in the clouds He does not describe Himself as descending to earth, nor as descending to the Mount of Olives. Many pre-tribulational any-moment rapture rescue promoters assert that the appearance of Jesus in the clouds, as described in the Olivet Discourse, is accompanied by His immediate earthly descent. Is that descent a descent to Armaggedon to battle, or is that descent to the Mount of Olives as the prophet Zechariah describes? Since the descent to earth of Jesus in those Olivet Discourse passages is not described by Jesus, pre-tribulational imaginations are entitled to create any scenario for the imaginary descent to earth they desire. This imaginary descent to earth in the Olivet passages is required for pre-tribulational any-moment rapture supporters to maintain their teachings. If there is no descent there of Jesus to earth, there is no pre-tribulational rapture. Unfortunately, no descent by Jesus is described in those Olivet passages, so any-moment pre-tribulational rapture promoters are obligated to manufacture one in order to maintain their theory. This is eisegesis, the reading into a text of a previous idea. Exegesis extracts from the text the meaning which is derived from the context. Inserting an earthly descent of Jesus into the Olivet Discourse description of the appearance of Messiah in the clouds, is eisegesis.

If pre-tribulational any-moment rapture promoters are correct, Jesus must return to earth in Matthew 24:30-31, despite the strange failure of any gospel writer to mention that significant detail in the Olivet Discourse. The strange fact that none of the gospel writers state that Jesus returns to earth in their descriptions of His heavenly appearance in the Olivet Discourse, can lend strong support to the conclusion that the earthly descent is not mentioned for a very significant reason: there is none at that time. These details present strong and compelling evidence for the conclusion that the heavenly appearance of Jesus in the Olivet Discourse is actually the non-pre-tribulational rapture. These details lend strong support to the conclusion that the heavenly appearance of Jesus in the Olivet Discourse is not the pre-millenial return to earth of Jesus, but that it is instead the rapture of the elect.

It seems very likely that the gospel writers who record Messiah's heavenly appearance in the Olivet Discourse fail to mention a descent to earth, because Jesus does not descend to earth at that time. The heavenly appearance of Jesus in the Olivet Discourse is actually the rapture rescue. The rapture rescue is another of a long list of signs which Jesus gave for the ending of the current Gentile age. The rapture is a sign that the Gentile Age is about to end, and that the descent of Jesus to earth to begin His thousand year reign is about to occur very soon. This is very likely why Jesus describes His own future heavenly appearance at the rapture as a "sign" in Matthew 24:30-31. Not only are plagues, famines, wars, and rumors of wars signs for the end of the age (Matt. 24:3), but the rapture itself is an additional sign which Jesus adds to the long list of signs for the end of this current age. This list of signs is also evidence for the soon beginning of the future kingdom age, which Judaism has been anticipating for centuries.

If Jesus was referring to the rapture when He described His heavenly appearance in the clouds to His disciples, then there are additional conclusions which can be drawn from the Olivet Discourse about events which will occur prior to the rapture. We can conclude that if Jesus was describing events mostly chronologically in the Olivet Discourse, then events listed prior to the apparent rapture descriptions must occur first, before the rapture.

One of the events Jesus lists prior to His heavenly appearance in the clouds, during His dialogue with the disciples on the Mount of Olives, is the event known as the "abomination of desolation." The description of the rapture occurs in Matthew 24:28, 30-31, but the "abomination of desolation" occurs in Matthew 24:15. It appears that these events are indeed given in chronological order. The desecration of the soon to be rebuilt Jewish sacrificial Temple will evidently occur before the snatching away from earth of Messiah's followers to be with Him. This is the same order of events which the Apostle Paul lists in II Thessalonians.

1 Now we beseech you, brethren, by the coming of our Lord Jesus Christ, and by our gathering together unto him,

2 That ye be not soon shaken in mind, or be troubled, neither by spirit, nor by word, nor by letter as from us, as that the day of Christ (the rapture) is at hand.

3 Let no man deceive you by any means: for that day shall not come, except there come a falling away first, and that man of sin be revealed, the son of perdition; (II Thessalonians 2:1-3)

Here the Apostle Paul refers to the rapture as "our gathering together unto Him." Then Paul specifically states that "that day" (the day of the rapture), will not occur unless the apostasy and the revealing of the man of sin occur first. The "man of sin" is the world's future evil pharaoh who will masquerade as Israel's Messiah.

Paul states in II Thessalonians 2:1-3 that a precondition for the day of the rapture ("our gathering together unto Him") is the prior revealing of the world's future evil pharaoh (the "man of sin").

Armed with this basic chronology, we can examine again the words of Jesus regarding the abomination of desolation in the Olivet Discourse, and compare it with Paul's description in II Thessalonians 2:1-3.

In Mark's version of the Olivet Discourse, Jesus refers to the abomination of desolation in chapter thirteen verse fourteen.

> *14 But when ye shall see the abomination of desolation, spoken of by Daniel the prophet, standing where it ought not, (let him that readeth understand,) then let them that be in Judaea flee to the mountains:*
>
> *15 And let him that is on the housetop not go down into the house, neither enter therein, to take any thing out of his house:*
>
> *16 And let him that is in the field not turn back again for to take up his garment.*
>
> *17 But woe to them that are with child, and to them that give suck in those days!*

18 And pray ye that your flight be not in the winter.

19 For in those days shall be affliction, such as was not from the beginning of the creation which God created unto this time, neither shall be.

20 And except that the Lord had shortened those days, no flesh should be saved: but for the elect's sake, whom he hath chosen, he hath shortened the days.

21 And then if any man shall say to you, Lo, here is Christ; or, lo, he is there; believe him not:

22 For false Christs and false prophets shall rise, and shall shew signs and wonders, to seduce, if it were possible, even the elect. (Mark 13:14-22)

Mark's gospel indicates that the immediate reaction to the abomination of desolation by those believers living in Judea should be immediate flight to the mountains. Later in Mark 13:26-27 we are also given an apparent description of the rapture. The appearance of Jesus occurs as He comes in the clouds (not to earth), and the gathering of believers apparently occurs supernaturally with the assistance of angels. Matthew is the only gospel which mentions the trumpet in the Olivet Discourse which occurs at the rapture.

31 And he shall send his angels with a great sound of a trumpet, and they shall gather together his elect from the four winds, from one end of heaven to the other. (Matthew 24:31)

This trumpet is probably the same trumpet of I Thessalonians 4:16-17.

16 For the Lord himself shall descend from heaven with a shout, with the voice of the archangel, and with the trump of God: and the dead in Christ shall rise first:

17 Then we which are alive and remain shall be caught up together with them in the clouds, to meet the Lord in the air: and so shall we ever be with the Lord. (I Thessalonians 4:16-17)

Most pre-tribulational teachers admit that in II Thessalonians 2:1 Paul is referring to the rapture. The very Greek word Paul uses to describe the rapture in II Thessalonians 2:1 is the same root Greek word used by Matthew in recording Messiah's description of the rapture in the Olivet Discourse (Matt. 24:31).

Paul refers to the rapture by the Greek word "episunagogees" (II Thess. 2:1), and Matthew translates the word Jesus used to describe the rapture gathering by use of the same root Greek word, although it is in a slightly different case ("episunaxousia"). By comparing the description of the rapture Paul gives in I Thess. 4:16-17, with the apparent description Jesus gives of His heavenly appearance in Matt. 24:30-31, we note the following similarities between both descriptions, which appear to detail the same event. Consider these parallels.

1. Both Jesus and Paul refer to a trumpet.
 • *4:16 For the Lord himself shall descend from heaven with a shout, with the voice of the archangel, and with the trump of God: and the dead in Christ shall rise first (I Thessalonians).*
 • *24:31 And he shall send his angels with a great sound of a trumpet, and they shall gather together his elect from the*

four winds, from one end of heaven to the other (Matthew).

2. Both Jesus and Paul refer to Jesus as coming.
• *4:16 For the Lord himself shall descend from heaven with a shout, with the voice of the archangel, and with the trump of God: and the dead in Christ shall rise first (I Thessalonians).*
• *24:30 And then shall appear the sign of the Son of man in heaven: and then shall all the tribes of the earth mourn, and they shall see the Son of man coming in the clouds of heaven with power and great glory (Matthew).*

3. Both Jesus and Paul describe the Messiah as in the "air," the "clouds," or the "heavens."
• *4:17 Then we which are alive and remain shall be caught up together with them in the clouds, to meet the Lord in the air: and so shall we ever be with the Lord (I Thessalonians).*
• *24:30 And then shall appear the sign of the Son of man in heaven: and then shall all the tribes of the earth mourn, and they shall see the Son of man coming in the clouds of heaven with power and great glory (Matthew).*

4. Both Jesus and Paul also describe the gathering of people as occurring in the "air," the "clouds," or the "heavens."
• *4:17 Then we which are alive and remain shall be caught up together with them in the clouds, to meet the Lord in the air: and so shall we ever be with the Lord (I Thessalonians).*
• *24:31 And he shall send his angels with a great sound of a trumpet, and they shall gather together his elect from the four winds, from one end of heaven to the other (Matthew).*

5. Both Jesus and Paul describe angelic involvement.
• *4:16 For the Lord himself shall descend from heaven*

with a shout, with the voice of the archangel, and with the trump of God: and the dead in Christ shall rise first (I Thessalonians).
• *24:31 And he shall send his angels with a great sound of a trumpet, and they shall gather together his elect from the four winds, from one end of heaven to the other (Matthew).*

6. Both Jesus and Paul describe only living people who are believers as being gathered (Jesus apparently refers to living believers as being gathered from the four winds of earth).
• *4:17 Then we which are alive and remain shall be caught up together with them in the clouds, to meet the Lord in the air: and so shall we ever be with the Lord (I Thessalonians).*
• *24:31 And he shall send his angels with a great sound of a trumpet, and they shall gather together his elect from the four winds, from one end of heaven to the other (Matthew).*

7. Both Jesus and Paul apparently describe a resurrection (Jesus apparently refers to a resurrection by referring to a gathering from the ends of heaven).
• *4:16 For the Lord himself shall descend from heaven with a shout, with the voice of the archangel, and with the trump of God: and the dead in Christ shall rise first (I Thessalonians).*
• *24:31 And he shall send his angels with a great sound of a trumpet, and they shall gather together his elect from the four winds, from one end of heaven to the other (Matthew).*

8. Both Jesus and Paul appear to be referring to a resurrection of only the elect.
• *4:16 For the Lord himself shall descend from heaven with a shout, with the voice of the archangel, and with the trump of God: and the dead in Christ shall rise*

first (I Thessalonians).

• *24:31 And he shall send his angels with a great sound of a trumpet, and they shall gather together his elect from the four winds, from one end of heaven to the other (Matthew).*

9. The trumpet in both descriptions by Jesus and Paul appears to belong to the Lord's side.

• *4:16 For the Lord himself shall descend from heaven with a shout, with the voice of the archangel, and with the trump of God: and the dead in Christ shall rise first (I Thessalonians).*

• *24:31 And he shall send his angels with a great sound of a trumpet, and they shall gather together his elect from the four winds, from one end of heaven to the other (Matthew).*

10. In both descriptions by Jesus and Paul the coming of Messiah is not described as a descent to touch the earth.

• *4:16 For the Lord himself shall descend from heaven with a shout, with the voice of the archangel, and with the trump of God: and the dead in Christ shall rise first (I Thessalonians).*

• *24:30 And then shall appear the sign of the Son of man in heaven: and then shall all the tribes of the earth mourn, and they shall see the Son of man coming in the clouds of heaven with power and great glory (Matthew).*

11. In both descriptions by Jesus and Paul there is no description of a white horse for Jesus to ride.

• *4:16 For the Lord himself shall descend from heaven with a shout, with the voice of the archangel, and with the trump of God: and the dead in Christ shall rise first (I Thessalonians).*

• *24:30 And then shall appear the sign of the Son of man in heaven: and then shall all the tribes of the earth mourn, and they shall see the Son of man coming in the clouds of heaven with power and great glory (Matthew).*

12. In both descriptions by Jesus and Paul there are no descriptions of an army accompanying Jesus.
 • *4:16 For the Lord himself shall descend from heaven with a shout, with the voice of the archangel, and with the trump of God: and the dead in Christ shall rise first (I Thessalonians).*
 • *24:30 And then shall appear the sign of the Son of man in heaven: and then shall all the tribes of the earth mourn, and they shall see the Son of man coming in the clouds of heaven with power and great glory (Matthew).*

These twelve parallels between two verses Jesus spoke and two verses Paul wrote seem to suggest they are evidently describing the same event. The above twelve similarities between Matthew 24:30-31 by Jesus and I Thessalonians 2:16-17 by Paul suggest that the ancient church leader John Chrysostom was justified, and probably correct, when he interpreted these passages by Jesus and Paul as being the same event. The evidence is so staggering, it appears absurd to interpret the Matthew 24:30-31 appearance of Jesus as the millenial return of Revelation 19:11 where Jesus rides a white horse accompanied by a heavenly army to fight the armies of earth.

It also appears absurd, from this staggering list of similarities, to interpret the Matthew 24:30-31 appearance of Jesus as the predicted Zechariah 14:4 touchdown to the Mount of Olives. The Biblical details suggest the Matthew 24:30-31 heavenly appearance of Jesus is none other than the predicted rapture of I Thessalonians 4:16-17.

Since it appears that it is most consistent to interpret Matthew 24:30-31 as the rapture, we can conclude from the chronology Jesus gives in the Olivet Discourse that the abomination of desolation precedes the rapture. Paul informs us of the same chronology in II Thessalonians 2:1-4 where he states that the revealing of the man of sin occurs before the rapture (before "our gathering together unto Him").

Paul may have used a gospel or circulating transcription of the Olivet Discourse for his rapture chronology in II Thessalonians 2:1-4 if he didn't get that information directly from the Lord Himself. Paul elaborates on information about the antichrist which doesn't appear in the gospels. The apostle seems to indicate the exact form in which the revealing of the future world pharaoh will occur.

> *3 Let no man deceive you by any means: for that day (the rapture) shall not come, except there come a falling away first, and that man of sin be revealed, the son of perdition;*
>
> *4 Who opposeth and exalteth himself above all that is called God, or that is worshipped; so that he as God sitteth in the temple of God, shewing himself that he is God. (II Thessalonian 2:3-4)*

The Apostle Paul seems to expressly indicate that the future evil pharaoh will be revealed by his act of entering the future rebuilt Jewish Temple of sacrifice and sitting there, displaying himself as the supreme deity. This is the form the abomination of desolation evidently takes at its inception. Like Lucifer, the future evil pharaoh will desire to assume the position of deity. It is this act of self-deification in Israel's future rebuilt sacrificial Temple which will apparently unveil or unmask the identity of that man of sin, earth's future world pharaoh. This unveiling is perhaps the very beginning of the abomination of desolation event predicted by Daniel the prophet (Dan. 9:27). Both Jesus and the Apostle Paul indicated that this act of abomination precedes the rapture.

This piece of chronological information is of strategic value in interpreting the book of Revelation. If we know the revealing of the false Messiah precedes the rapture, then we know that events in the book of Revelation preceding this identity revealing of the antichrist will occur before the rapture. The question left to be answered is, "When is the future evil world pharaoh's identity revealed to the world in the chronology of the book of Revelation?" If we can identify the moment of his public self-exaltation in the future sacrificial Temple, we can then determine which events precede the abomination of desolation and determine which book of Revelation events precede the rapture rescue of Messiah's followers.

When is the identity of the "man of sin" revealed to the world? Where does the abomination of desolation occur in the book of Revelation?

I have been unable to locate the exact moment of the future world pharaoh's future self-exaltation in the book of Revelation. The exact moment of the abomination of desolation seems to be hidden in the book of Revelation, but there are specific clues which indicate when it could not happen.

In Revelation 11:7 the man of sin is referred to as "the beast that ascendeth out of the pit." This description is of paramount value, because it indicates that the man of sin receives his "demonic anointing" from a spirit which has been restrained in the bottomless pit (II Thess. 2:6-8).

When this spirit demon, or fallen angel, is released from the bottomless pit, the demonic anointing of the false Messiah suddenly becomes an option.

The Apostle John in his apocalyptic vision refers to a moment in which the bottomless pit is opened. In Revelation 9:1-4, at the fifth trumpet judgment, demonic locusts ascend out of the bottomless hole to torment human beings who have not been sealed by the Lord. This is the only mention of the release of demons from the horrible abyss mentioned in the book of Revelation, other than when Satan is loosed from that same pit after being chained for 1,000 years during the millenial reign of Jesus on earth (Rev. 20:3).

Since it is not permissible to add to the book of Revelation, we are forced to conclude that the opening of the bottomless pit at the fifth trumpet judgment in Revelation 9:1-3 is the only event in the book of Revelation where the fallen "beast" demon which "anoints" or possesses the man of sin can possibly ascend out of the pit. It seems likely that the future evil world pharaoh cannot receive his "beast" anointing through demon possession until Revelation 9:1-3 occurs. This would indicate that most of the events in the book of Revelation preceding Revelation chapter nine occur before the rapture. This assumption seems to be validated by the events Jesus Himself predicted as transpiring prior to His heavenly appearance at the rapture in the Olivet Discourse.

Plagues Precede the Exodus
It has been noted that Jesus apparently describes the rapture in Matthew 24:30-31, because of at least twelve similarities to I Thessalonians 4:16-17. In this Olivet Discourse description Jesus lists the rapture as a "sign." It is a sign

the Gentile age is about to end, and that the millenial kingdom is about to begin. This rapture "sign" is preceded by numerous other signs which Jesus lists in detail. One of the events preceding the Olivet Discourse rapture (Matt. 24:30-31) is the abomination of desolation (Matt. 24:15). The plagues, famines, pestilences, and celestial events which Jesus also lists as signs in the Olivet Discourse, precede chronologically both the rapture, and the abomination of desolation as described by Jesus. What is interesting is the fact that these preliminary signs in the Olivet Discourse which occur before the rapture and before the abomination of desolation have specific parallels to the book of Revelation.

Revelation 9:1-3 brings us to the opening of the bottomless pit. It is at this time the demon which inhabits the future antichrist probably ascends out of this abyss (Rev. 11:7). Since the abomination of desolation occurs before the rapture, and the abomination cannot occur until the demon inhabiting the world pharaoh is released from the bottomless pit, we can conclude that generally everything before Revelation 9:1-3 occurs before the abomination of desolation and before the rapture.

The events in the book of Revelation preceding the beast demon's ascent from the bottomless pit have amazing parallels to the plagues, famines, pestilences, and celestial signs predicted by Jesus in the Olivet Discourse. Let us examine some of them.

First, in Matthew 24:6 Jesus predicted that before He appeared in the clouds (at the rapture) there would be wars and rumors of wars. While this phenomenon has always been with us, in Revelation 6:4 the red horseman, who is released at the second seal, takes peace from the earth, and the horseman is given a great sword. The taking of peace from the earth implies war.

Secondly, in Matthew 24:7 Jesus predicted famines would occur before the rapture. Famines are loosed in Revelation 6:5-6 at the third seal with the appearance of the black horseman holding a pair of balances in his hand. A voice announces, "A measure of wheat for a penny, and three measures of barley for a penny, and see thou hurt not the oil and the wine."

Third, in Matthew 24:7 Jesus speaks of pestilences which would occur before the rapture. In Revelation 6:7-8, the pale horseman brings death to one quarter of the earth's population by the sword, by hunger, by death, and maybe through diseases affecting the beasts of the earth (Van Impe, 1982, p. 78).

Fourth, in Matthew 24:9 Jesus predicts what may be universal persecution against His followers saying, "Then shall they deliver you up to be afflicted, and shall kill you: and ye shall be hated of all nations for my name's sake." At the fifth seal in Revelation 6:10-11 martyrs are seen under the heavenly altar.

Fifth, in Matthew 24:29 Jesus predicts events before the rapture which appear to occur at the sixth seal in Revelation 6:12-13. Jesus predicted the sun would be darkened, the moon would not give its light, the stars would fall from heaven, and the powers of the heavens would be shaken. The sixth seal unlooses these very events.

Sixth, in Matthew 24:27 Jesus predicted that before the rapture earthquakes would occur in diverse places. In Revelation 6:12, 14 an earthquake with worldwide consequences affects perhaps all regions of the earth at the sixth seal.

Finally, in Matthew 24:22 Jesus said, "And except those days should be shortened, there should no flesh be saved: but for the elect's sake those days shall be shortened." In Revelation 8:12, at the fourth trumpet, celestial day and night lights are evidently shortened by one third.

These seven predictions which Jesus made to His disciples during His dialogue with them on the Mount of Olives, as we have seen, have parallels to the book of Revelation prior to the opening of the bottomless pit. The bottomless pit releases the demon spirit which empowers the man of sin, the future world pharaoh. The revealing of the identity of this demon-empowered man through his self-exaltation in the sacrificial Temple is the beginning of the abomination of desolation. The abomination of desolation occurs before the rapture Jesus predicted (Matt. 24:28, 30), and before the rapture Paul predicted (II Thess. 2:1-3). These details indicate that incidents prior to the opening of the bottomless pit will precede the rapture in the book of Revelation. Those events prior to the opening of the bottomless pit occur during the first part of the seven years in the book of Revelation.

Future Plagues Parallel the Plagues on Egypt
As just seen, Jesus predicted devastating calamities for the world before His appearance in the clouds at the rapture (Matt. 24:30-31). We have also seen that these world calamities Jesus predicted are parallel to events listed in the book of Revelation before the bottomless pit is opened. At the opening of the bottomless pit in Revelation. 9:1-3, the demon which empowers the future world pharaoh is finally loosed to begin its activity after having been restrained for

centuries (II Thess. 2:6-7). We have noted that both the Apostle Paul and Jesus predicted that the abomination of desolation will precede the rapture. The "abomination of desolation," we have further seen, is evidently initiated the moment the future world pharaoh attempts to exalt himself in Israel's rebuilt, sacrificial temple. Since this self-deification in Israel's future Temple of sacrifice precedes the rapture, according to both Paul and Jesus, we can conclude that the events in the book of Revelation before the releasing of the demon which ascends out of the bottomless pit to anoint the counterfeit Messiah will occur before the rapture.

Generally speaking, the events before the opening of the bottomless pit in Revelation 9:1-3 apparently occur before the rapture. Those events not only have parallels to the calamities Jesus predicted in the Olivet Discourse, but those book of Revelation events also have parallels to the plagues which occurred in Egypt just before Israel's exodus. Before the mass exodus of the Messiah's true followers to heaven at the rapture to escape earth's future pharaoh, it appears that plagues will come upon the earth which will have parallels to the plagues that came upon Egypt before Israel's escape from the historical pharaoh. These plagues will precede the rapture, just as the plagues on Egypt preceded the exodus. The future deliverance of the Lord's followers will be from earth's future evil pharaoh, just as the historical deliverance of Israel was from ancient Egypt's evil pharaoh.

Before the deliverance of Israel from Egypt, ten plagues came upon pharaoh's land at the hands of two witnesses, Aaron and Moses. The plagues which they brought upon Egypt are strikingly similar to the plagues in the book of Revelation which precede the rapture. It has been suggested by some Biblical commentators that some of the plagues recorded in the book of Revelation come at the hands of two of the Lord's witnesses. In Revelation chapter eleven we read of two men having a three and one-half year miracle ministry which is strongly reminiscent of Moses and Aaron who stood before Egypt's pharaoh. Notice the following interesting statement about these two future prophets.

> *6 These have power to shut heaven, that it rain not in the days of their prophecy: and have power over waters to turn them to blood, and to smite the earth with all plagues, as often as they will. (Revelation 11:6)*

The amazing parallel here to ancient Egypt is that the first plague which came upon Egypt came at the hands of Moses and Aaron when they also turned the water into blood.

> *19 And the LORD spake unto Moses, Say unto Aaron, Take thy rod, and stretch out thine hand upon the waters of Egypt, upon their streams, upon their rivers, and upon their ponds, and upon all their pools of water, that they may become blood; and that there may be blood throughout all the land of Egypt, both in vessels of wood, and in vessels of stone.*
>
> *20 And Moses and Aaron did so, as the LORD commanded; and he lifted up the rod, and smote the waters that were in the river, in the sight of Pharaoh, and in the sight of his servants; and all the waters that were in the river were turned to blood.*
>
> *21 And the fish that was in the river died; and the*

> *river stank, and the Egyptians could not drink of the water of the river; and there was blood throughout all the land of Egypt. (Exodus 7:19-21)*

This first plague at the hands of Moses and Aaron is strangely parallel to the plague which occurs at the second trumpet in Revelation 8:8-9.

> *8 And the second angel sounded, and as it were a great mountain burning with fire was cast into the sea: and the third part of the sea became blood;*
>
> *9 And the third part of the creatures which were in the sea, and had life, died; and the third part of the ships were destroyed. (Revelation 8:8-9)*

Does this plague come at the hands of the two Revelation chapter eleven prophets who have the power to turn water to blood?

There are other plagues in the book of Revelation before and during the opening of the bottomless pit which parallel the plagues on ancient Egypt. There is hail (Rev. 8:7), there are locusts (Rev. 9:3-11), and there is darkness (Rev. 6:12). All these Egyptian like plagues from Revelation 6–9:10 occur before the abomination of desolation, which, as we have seen, precedes the rapture, the future mass exodus from earth to heaven of both living and resurrected followers of the Lord. These future plagues, as was true with ancient Egypt, are directed against those who don't know the Lord, who are not sealed. Those who know the Lord are directed by Jesus to pray that they may be preserved through the horrible calamities which are to come upon the world.

36 Watch ye therefore, and pray always, that ye may be accounted worthy to escape all these things that shall come to pass, and to stand before the Son of man. (Luke 21:36)

It is possible for the Lord's people to be preserved through these future horrible plagues, because the Scriptures record that the Israelites were also preserved when the plagues on ancient Egypt were poured out. The plagues on ancient Egypt are evidently a prototype blueprint of calamities which precede the future rapture rescue exodus predicted by Jesus and the Apostle Paul (Matt. 24:28, 30-31; I Thess. 4:16-17). Believers should pray that they may be accounted worthy to be preserved through the many calamities which are going to come upon the world (Luke 21:36).

Chapter Nine

No Intervening Events Required?

The trumpet sounded. Then it sounded again. This time the trumpet seemed to get closer! The assembled crowd rushed out in anticipation for the rapture and the coming of the Messiah in the clouds. When they got outside their excitement turned to disillusionment. The trumpet was being sounded by an alcoholic (Chandler, 1993, p. 83).

William Miller was a Baptist lay preacher who had become obsessed with the idea of calculating the time of the rapture. He had originally set a date for 1843. When that calculation proved to be in error, revisions were made to the theory and a new date was settled upon with some of the people who had climbed aboard his movement. The new date was corrected to 1844. Astoundingly, a new wave of anticipators swept into and were added to the numerous tens of thousands of watchers who had already climbed aboard Miller's movement. Unfortunately, the new date which had been set for the rapture proved to be as erroneous as the earlier calculation.

William Miller was a man who had forsaken his childhood religious rearing and had taken a path into deism. At about the age of thirty-four he had joined the Baptists through a revival. He became interested in Biblical prophecy and chronological calculations for the end of the world. His passion for Biblical prophecy led to some unfortunate, unwise, and unbiblical conclusions about the specific time for the Messiah's coming in the clouds for His followers. Specifically, Miller began to subscribe to mathematical theories which he found he was unable to keep to himself. His opinions were transformed into public proclamations of anticipation for a date of the rapture which was calculated to occur in 1843.

There were some unfortunate incidents to this whole episode. People had given away property and had sold things. Others faced mocking and ridicule as a result of their previous gullible behavior. Stories of bizarre incidents have been preserved about some of the people participating in the watch for Messiah, while probably most retained their composure through the whole William Miller affair.

Not all of Miller's ideas about prophecy were unbiblical. Much of what Miller had to say came from the Scriptures. Unfortunately, his false ideas created an embarrassing situation for all churches and religious groups which shared some of his basic Biblical beliefs. Others who believed in the same literal approach to prophecy Miller practiced were chastened, if not totally embarrassed, by William Miller's presumptuous mistakes. It was during the aftermath of the Millerite debacle that the doctrine of an *any-moment rapture, without any required intervening events,* suddenly seemed to catch hold. There were probably many churches, believers, denominations, and religious organizations which may have shared some of the basic fundamental values Miller had believed. These fundamental values are what may have given William Miller such credibility with so many thousands of people. These other religious groups were determined, most likely, to learn from the mistakes of others. They probably became determined not to repeat the errors of William Miller.

The move to prevent future William Millers seems to have taken hold of many religious organizations which have come down to us from those years, and has been latched onto by other fledgling groups which were formed later, but which also were sure they didn't want to repeat the errors of the past. The development of the *any-moment rapture requiring no intervening events* was a peripheral development of John Nelson Darby's rapture teachings. The doctrine of an *any-moment rapture which requires no intervening events* has been christened as the teaching of "imminency." Imminency was a post-Miller doctrine which apparently started out as a concept urging watchfulness for the coming of Messiah. Watchfulness is certainly a Biblical teaching and a command which was given by Jesus Himself. When the doctrine of watchfulness was eventually processed through a committee, it was elaborated upon and came to be endowed with the specific enunciations it holds today. Evidently the concern of some committee members may have become the idea of preventing the kind of watchfulness which William Miller had erroneously promoted. The watchfulness doctrine was transformed by the committee into an anti-date setting formula. That formula would prevent William Miller types from setting dates for the rapture.

"Imminency," as it is now termed, specifically is defined with the idea that no intervening events are required for the rapture to occur. The idea is clearly taught in the idea of imminency that the rapture is an event which can happen at any-moment, even in the next second or two. The teaching appears to be highly motivated at discouraging any potential William Miller look-a-like reenactments. Among organizations adopting and enforcing the statement, the teaching has clearly succeeded quite admirably in protecting the dignity of those groups from being embarrassed by potential William Miller styled rapture date setters, because by enforcing the imminency doctrine, date setters are prohibited from membership in these organizations. Imminency is a powerful anti-date setting formula.

Is the teaching of "imminency" Biblical? The teaching of imminency appears to have its origins in Scripture. The idea of watching for the coming of the Messiah is clearly a Biblical command. The command to watch was given by Jesus Himself. A survey of the context in which the command to watch was given even states that there is uncertainty involved in the time of the coming of the Lord. This leads most people to conclude that the teaching which is called "imminency" is of clearly Biblical origin. The doctrine of "imminency" also discourages the erroneous behavior which characterized the Millerite movement, the practice of setting dates for the rapture.

It has been documented down through history that when some groups attempt to correct errors, there is often the tendency to overreact in dealing with an issue or problem which is observed. The story of history is that many organizations overcorrect a matter when dealing with a situation. Sometimes in correcting an error, the tendency is to misjudge the solution, and in the process of creating a remedy, errors as bad as the original can be made which create a totally new set of problems. This appears to be the situation which has resulted in the rush to prevent any potential William Millers who might be lurking in the shadows of churches and religious groups which don't want to be contaminated by similar Miller type happenings in the future.

There is solid Biblical evidence which indicates there were intervening events which had to occur before the rapture could take place. These Biblical events are documented in the Scriptures. We have already reviewed Biblical predictions which seem to indicate certain events must transpire before the rapture. If the Scriptures do predict that intervening events must occur before the rapture, then the idea of "imminency" must be called into question. Imminency states that no intervening events are required for the rapture to occur. The question is, how accurate and Biblical is the concept which states *no intervening events are required for the rapture to occur?* Is the idea that no intervening events are required for the rapture to occur a Biblical idea? If that teaching is accurate, has it always been a true statement? Were there times in history when the rapture couldn't occur? Was there a certain stage in history at which time the rapture suddenly became an event which no longer had any intervening events? In other words we can ask, "If imminency is correct now, was there a time when imminency was not a correct teaching?" One could also ask, "Is there a time when imminency became a correct teaching?"

Why should anyone be concerned about the issue of "imminency?" What difference does it make if Jesus can rapture His followers at any-moment or if He can't rapture His followers at any-moment? To answer that, it would probably be good to use the words someone else has used: if you don't stand for something, you'll fall for anything. If the teaching of imminency is correct (that is, that there are no intervening events required for the rapture to occur), then the preceding chapters of this book can be labeled as heresy and false teaching. If it is Biblical to say that "no intervening events are required for the rapture to occur," then the idea of a rapture having signs is just nonsense.

"Imminency" teaches that no intervening signs are required to occur before the rapture, and, consequently, a sign-less rapture is supposedly Biblical. If one cannot watch for rapture signs, looking for rapture signs is a waste of time.

There is another practical reason for being concerned with the teaching of "imminency." Accepting or rejecting this doctrine is grounds for acceptance or rejection of an individual into membership among many church and religious organizations. These organizations may be particularly concerned with members potentially reenacting William Miller's date setting errors, and they don't want to be threatened with the stigma of some false teacher in their association.

There are very practical reasons, consequently, for examining the "imminency" rapture teaching to determine how Biblical and accurate it may be. It is very practical to determine whether the idea of *a rapture that can happen at any-moment without any intervening events being required* is indeed Biblical, or possibly in error. If the teaching is in error, where is the error, and how can it be corrected? It is worthwhile to examine the *any-moment, imminent rapture teaching* to determine if it is as orthodox and Biblical as its advocates claim.

Imminency as enunciated by the phrase, "No intervening events required," is a recent doctrine. Watchfulness is a Biblical teaching, but the idea that intervening events cannot postpone a rapture rescue is historically non-existent. That section of the "imminency" idea is brand new, and has no detailed history in the development of church doctrine through the centuries. Does that make it wrong? No. The authority for a Biblical teaching is not church history or tradition, as helpful as those things may be, rather, the Bible itself is the fundamental basis for authority and for determining the correctness or incorrectness of a teaching.

While the Biblical support for a teaching is of primary value to real believers, sometimes the historical development of a teaching provides illumination which can help resolve tension. We have already examined the historical development of the "no intervening events" aspect of "imminency." Its roots are evidently grounded in the William Miller events of the nineteenth century. The imminency teaching was also an apparently peripheral development from Darby's pre-tribulational teachings about the rapture. These two events are of primary historical significance in examining the idea of an *any-moment, imminent rapture.* As the stigma of Miller's mathematical calculations for the date of the rapture lingered over the nineteenth century, Darby's teachings about the rapture seemed to fill the vacuum which had been created and provided a new respectability for the Biblical passages about the rapture rescue of believers which had been so traumatized and stained by William Miller's misguided mathematical calculations.

The imminency doctrine, because of its enforcement by religious organizations, has saved those religious groups adopting and enforcing it much embarrassment. At the same time, some peculiar and strange theological developments have occasionally resulted. One peculiar teaching which has resulted was the hybrid idea of a post-tribulational rapture which was imminent (it could happen at any-moment, despite the fact the rapture would supposedly transpire at the end of the seven years of turmoil in the book of Revelation). With this strange oddity in theology, it would not be unusual for an imminent "mid-tribulational rapture" teaching or an imminent "pre-wrath" rapture teaching to soon make their appearances on the horizon. For some organizations, the only critical issue is "imminency." The idea is that as long as an individual agrees to the idea that the rapture can happen at any-moment, even in the next second, it does not matter whether one tries to place the rapture at the end, or somewhere in the middle of the seven years of turmoil. With an imminency view, one could suppose we might be going through the seven years in the book of Revelation right now and not even know it!

Those individuals who believe that a pre-tribulational rapture rescue of believers is incorrect, but who wish to work with an organization which has adopted the anti-Millerite solution of an *any-moment rapture without any intervening signs,* may be forced to unusual compromises. They may not wish to abandon their non-pre-tribulational understanding of the rapture which they have derived from the Scriptures, but they may also want to work with an organization which has an anti-Millerite formula. The solution for those folks is to either keep silent about their rapture beliefs, or to come to some kind of ideological compromise which allows mutually exclusive ideas to coexist. This appears to account for the unusual oddity of an "imminent post-tribulational rapture."

Imminency, the idea of an *any-moment rapture with no intervening events required*, logically leads to pre-tribulationalism. The probable reason for this is that the development of the detailed explanation for imminency took place in a committee heavily influenced by Darby's pre-tribulational rapture views. There are other possible solutions to counteract Millerite date setting tendencies, but many organizations (some perhaps unknowingly) are locked into the anti-date setting solution of "imminency." Imminency is, at its foundation, a logical precursor to a pre-tribulational rapture. Imminency is a logical precursor to pre-tribulational teaching because imminency is a peripheral doctrine heavily influenced by the pre-tribulational theories of John Nelson Darby.

While imminency advocates will undoubtedly defend the idea of imminency as a completely Biblical doctrine derived from the Scriptures, the claim is fragmented by the inability of different pre-tribulational rapture groups to agree on the Biblical foundation for the teaching of imminency. Imminency supporters have trouble agreeing on which Scriptures actually teach imminency. It is difficult to exegetically analyze and critique varying positions on imminency which are all supposedly supported by the Bible, but by different Scriptures. One set of Scriptures is rejected as valid by one group, and so to analyze those Scriptures results in the evaluation by others that they are not valid passages of concern.

It can be exasperating to evaluate imminency rapture positions because of these varying ideas about which Scripture passages provide valid support for imminency. Some supposed verses for imminency are thought to even contradict the idea of imminency for another group. Because of this paradoxical situation, the attempt will be made to present Biblical evidence which indicates that an *unconditionally imminent, any-moment rapture requiring no intervening events* was not a valid truth, at least at some points in the Biblical record.

The Power Trip
Just before Jesus ascended into heaven, He gave His disciples some instructions to carry out after He was gone.

> 4 And, being assembled together with them, commanded them that they should not depart from Jerusalem, but wait for the promise of the Father, which, saith he, ye have heard of me.
>
> 5 For John truly baptized with water; but ye shall be baptized with the Holy Ghost not many days hence.
> (Acts 1:4-5)

The idea of unconditional imminency, *a rapture which can happen at any-moment without any intervening events,* is contradicted by the words of Jesus Himself. Here in Acts, Jesus clearly orders His disciples to carry out His instructions. The command that Jesus gave them was that they were to go to Jerusalem and wait there for the promise of the Father. Accompanying this command was a specific prediction by Jesus declaring that the disciples would be rewarded through their obedience with the baptism of the Holy Spirit.

Many believe that after the resurrection, Jesus was no longer ignorant about the specific time for the rapture. Jesus, in His incarnation prior to the resurrection, had limited Himself concerning His divine attributes. The limit on Messiah's omniscience (all-knowingness) is seen in the Olivet Discourse, where Jesus specifically claims ignorance concerning the time of His own return.

> 36 But of that day and hour knoweth no man, no, not the angels of heaven, but my Father only.
> (Matthew 24:36)

Is it possible that after the resurrection, Jesus was no longer limited by a lack of knowledge concerning the time of His own return? However one may answer that question, the clear idea Jesus conveyed to His disciples was that they were to take a trip and wait in Jerusalem. Jesus assured them with a clear prediction that their obedience would be rewarded by their reception of the baptism with the Holy Spirit.

The promise of the Father is a reference to Joel 2:28, 29. The promise of the Father, Jesus explained, would involve the reception of power.

> *8 But ye shall receive power, after that the Holy Ghost is come upon you: (Acts 1:8)*

The prophet Joel explained the promise of the Father as something which would come upon all flesh.

> *28 And it shall come to pass afterward, that I will pour out my spirit upon all flesh; and your sons and your daughters shall prophesy, your old men shall dream dreams, your young men shall see visions:*
>
> *29 And also upon the servants and upon the handmaids in those days will I pour out my spirit. (Joel 2:28-29)*

The promised power is continuing today as clearly seen by the supernatural wonders demonstrated worldwide which accompany those who believe the Scriptures. Truly, all flesh are today participating in this Holy Spirit power outpouring, for the outpouring is worldwide in scope.

There was a purpose to the Father's promise. Jesus inferred that the Father's promise was an enduement of power which would help them to be witnesses.

Jesus indicated that the disciples would be aided by this Holy Spirit power enduement as they went throughout the world. The question is this: *could the rapture have occurred during this waiting period for promised power?*

If the rapture had occurred before the disciples had received the promised power of the Holy Spirit, would Jesus have been turned into a liar? Jesus had told His disciples to wait for the promise of the Father, assuring them that they would be rewarded. The clear prediction by Jesus was that what the Father promised would surely come to pass in the lives of the disciples.

Could the rapture have occurred before the disciples received the promise of the Father? The conclusive answer which we can come to, is that the teaching of an *any-moment rapture* was incorrect before the power enduement was received by the disciples.

> *8 But ye shall receive power, after that the Holy Ghost is come upon you: and ye shall be witnesses unto me both in Jerusalem, and in all Judaea, and in Samaria, and unto the uttermost part of the earth. (Acts 1:8)*

There are serious problems with the teaching of an *imminent, any-moment rapture requiring no intervening events.* The first major problem that presents itself, is that the Acts 1:4-5, 8 Scriptures indicate that a doctrine of an *unconditional, imminent, any-moment rapture,* was not theologically possible before the reception of the Father's promise of power by the disciples. The receipt of the Father's promise happened to occur ten days later on the day of Pentecost. The Pentecostal outpouring of power had to occur before the rapture to prevent Jesus from becoming, at the very least, a liar or false prophet. Because Jesus is God, Jesus could not lie. Evidently, if the idea of an *any-moment rapture requiring no intervening events* is a Biblical doctrine, it certainly was not a theologically correct doctrine during the wait by the disciples for the promise of the Father.

An Expanding Prophecy
There are further problems which present themselves to the honest inquirer who wishes to determine if an *imminent, any-moment rapture requiring no intervening events* is really a Biblical teaching. There are other details which present themselves in the section of Scripture where Jesus promised His disciples they would receive the Father's promise (which would be a power enduement). In explaining to His disciples the power enduement, which was the promise of the Father, Jesus implied that a strategy for evangelism had been established which had to be executed prior to the rapture. This evangelism strategy had multiple parts, and each part had to be participated in by the disciples and Apostles themselves. The promise of a power enduement included Messiah's evangelism strategy, with several sections to the strategy: Jerusalem, Judea, Samaria, and the uttermost part of earth.

> 8 *But ye shall receive power, after that the Holy Ghost is come upon you: and ye shall be witnesses unto me both in Jerusalem, and in all Judaea, and in Samaria, and unto the uttermost part of the earth. (Acts 1:8)*

The evangelism plan which Jesus outlined included, first of all, a witness to the theological heart of Israel, Jerusalem. Secondly, the evangelism plan would include the surrounding area of Judea. Thirdly, the evangelism strategy would expand to encompass the area known as Samaria. Fourth, the evangelism strategy for the disciples would be extended to include the uttermost part of the earth.

Messiah's outline for the evangelism plan disclosed to His disciples clearly predicts that the disciples would successfully enter all phases of that entire plan. If the disciples failed to enter all parts of the outlined strategy, Jesus at the very least would have been proven to be a liar or a false prophet. If the rapture occurred before the evangelism plan could be entered in all of its phases, by at least some of the disciples and Apostles to whom Jesus spoke, what Jesus predicted as happening would not have happened.

The clear implication is, that the scope of world evangelism by, at the very least, Messiah's immediate disciples, would take precedence over any possibility of the rapture. If the rapture was possible during the lives of the disciples, one must logically conclude it was possible only after they had successfully entered the final phase of the outlined plan for evangelism which Jesus had outlined for them. The final phase would be the mission of being witnesses to the "uttermost part" of the earth.

Concentric Circles
Some who have analyzed the plan of evangelism revealed by Jesus to His disciples have concluded that the plan can be represented by a series of concentric circles. The inner circle in Messiah's evangelism plan was Jerusalem. The rapture was not a possible event while the disciples remained in Jerusalem. The stay in Jerusalem was part of the plan, to be sure, but it was only the innermost circle on the plan.

Only the first phase of the strategy which Jesus had disclosed to His disciples would have been fulfilled if they had been raptured while they were still in Jerusalem. The clear implication is that the disciples could not logically consider an *any-moment rapture with no intervening events* as a logical or legitimate hope while they were occupying the first inner circle in the worldwide strategy for evangelism. If Jesus had told the truth, the disciples had other phases of the evangelism strategy to fulfill. The whole purpose of the power enduement appears to be wrapped up in this strategy for evangelism. The disciples had received the promise of the Father so that they could be effective witnesses. The first phase required this power enduement. The fact that other phases had not yet been entered by the disciples meant that there would necessarily be a delay in the possibility of an actual rapture.

As the disciples moved into the second phase of Messiah's evangelism strategy, the witness to Judea, the disciples could feel confident they were in the will of the Lord. The disciples needed the power enduement for the second phase of the strategy as well. If the rapture had occurred in the second phase of Messiah's plan of evangelism, the rapture would have prevented any of the succeeding phases from being fulfilled by the disciples. If the rapture had occurred after the disciples reached Judea, the second concentric circle would have been entered in the evangelism strategy, but the other phases would have gone uncompleted by the very disciples to whom Jesus had predicted successful accomplishment of all phases of His evangelism plan. Realistically, if the rapture had occurred while the disciples were in the Judean phase of the plan, Messiah's prediction that the disciples would fulfill His complete evangelism strategy would have proven to be false. The rapture was not theologically possible while the disciples were in the Judean phase of world evangelism.

The third phase of Messiah's world evangelism strategy for His disciples and Apostles was Samaria. Samaria has been pictured as the third concentric circle in the global strategy of world evangelism. Samaria was a semi-Gentile region of ethnic culture. Samaria's population was the product of interracial marriages by Jews and Gentiles. By reaching Samaria, the bringing of the gospel to the semi-Gentile races had begun. Samaria represents the very threshold of the Gentile world. Samaria was involved in Messiah's global strategy of evangelism, because God not only loved the Jews and Israel, He loved the semi-Jewish/semi-Gentile races as well. Samaria demonstrated Messiah's concern and love for even the semi-Jews of this planet. The Scriptures below show fulfillment of Messiah's predictions that the disciples and Apostles would also reach Samaria.

> *14 Now when the apostles which were at Jerusalem heard that Samaria had received the word of God, they sent unto them Peter and John:*
>
> *15 Who, when they were come down, prayed for them, that they might receive the Holy Ghost:*
>
> *16 (For as yet he was fallen upon none of them: only they were baptized in the name of the Lord Jesus.)*
>
> *17 Then laid they their hands on them, and they received the Holy Ghost. (Acts 8:14-17)*

Was the rapture possible when the apostles and disciples had finally reached Samaria? If the rapture had occurred while the disciples and Apostles had only entered the third phase of Messiah's evangelism strategy, the fourth phase of Messiah's evangelism strategy would have not been entered, and would not even have begun to be fulfilled by the very disciples and Apostles who had been assured by Jesus that they would successfully enter the fourth phase. At the very least, if the rapture had occurred in the third phase of the evangelism strategy, Jesus would have been proven to be a liar, or a false prophet. Realistically, the rapture was not even theologically possible while only phase three of the global evangelism strategy had begun. The disciples and Apostles needed the power enduement for the third phase of the plan, as much as they needed that power enduement for the first phase of the plan.

The third phase of the world strategy for evangelism demonstrated God's love for semi-Jews/semi-Gentiles, but the fourth and final phase was required to demonstrate God's love for the Gentiles. The fourth phase also had to be entered by the disciples and Apostles to whom Jesus spoke to demonstrate that Jesus was not a liar or a false prophet.

The fourth concentric circle in Messiah's global strategy for evangelism has been defined by the phrase "uttermost part of the earth." The fourth phase proves John 3:16 is correct. The fourth phase proves that God loves Gentiles as well as semi-Jews and as well as Israel. It is only after the Apostles and disciples had entered the fourth phase of Messiah's global strategy for evangelism that we can begin to entertain the notion of a possible rapture. The enduement of power was needed by the Apostles and disciples for effective witness to the Gentiles, as much as the power enduement was needed for effective witness to the Jews in Jerusalem. It is only when the Apostles and disciples entered the fourth phase of Messiah's global strategy for evangelism that we can finally see the fulfillment of Messiah's assurance that the disciples would participate in all phases of God's global strategy for evangelism. The participation of the Apostles and disciples to whom Jesus spoke in the fourth concentric circle of the global strategy proved that Jesus had correctly predicted their participation in all phases of the plan for world evangelism. It also proved God's love for all people groups.

> 8 But ye shall receive power, after that the Holy Ghost is come upon you: and ye shall be witnesses unto me both in Jerusalem, and in all Judaea, and in Samaria, and unto the uttermost part of the earth. (Acts 1:8)

Evangelism and the Rapture

Did the entrance of the disciples and Apostles into the fourth phase of global evangelism make the teaching of an *any-moment, imminent rapture requiring no intervening events* possible? Jesus had revealed that the four phases for the global strategy of evangelism would be completely participated in by at least some of the Apostles and disciples to whom He spoke in Acts chapter one. That prediction had to be fulfilled. If the rapture had occurred before any of the disciples or Apostles had participated in all four phases of the evangelism strategy, Jesus would have been shown to be a liar or false prophet.

After all four phases had been participated in by some of the disciples and Apostles to whom Jesus had revealed the plan, was an *unconditional, any-moment rapture requiring no intervening events* possible? Could the rapture occur then at any-moment?

Messiah's global plan for evangelism, as revealed to His disciples and Apostles in Acts chapter one, appears to correspond to a specific prediction Jesus made in the Olivet Discourse.

14 And this gospel of the kingdom shall be preached in all the world for a witness unto all nations; and then shall the end come. (Matthew 24:14)

This prediction by Jesus to the disciples during the Olivet Discourse appears immediately before the abomination of desolation in Matthew's gospel. A similar verse appears much earlier in Mark. The word for "nations" in Greek is translated "ethnesin." You may be able to recognize our word "ethnic" in it. The possible inference in the Olivet Discourse is that complete world evangelism may be accomplished prior to the abomination of desolation. It is possible this verse indicates only that complete world evangelism will be accomplished by the end of the Gentile era, by the end of the seven years listed in the book of Revelation. If the latter meaning is intended by Jesus, it would appear that completed world evangelization may occur in the time remaining after the rapture, which would imply the bulk of the complete work of world evangelization will have been completed prior to the rapture with only a slight "mop up job" for completely finishing the task taking place after the rapture.

There are evangelism organizations today which are involved in reaching the people groups of earth by systematic methods. Some of these organizations will list people groups which are yet unreached. Recently news about a Bible translation group was publicized in which they estimated the task of translating the Scriptures for all of the remaining unreached people groups of the world would take about fifty years for them to accomplish. Is it possible that world evangelism to every ethnic group on earth is a required event for the rapture to occur? If Jesus mentioned the rapture in the Olivet Discourse, which seems almost beyond doubt from the evidence we have already examined, and if the Olivet Discourse is listed in chronological order, it would appear that Jesus Himself might have predicted world evangelism as a prerequisite for the rapture, but certainly as a prerequisite for the end of the current Gentile era, which ends in less than four years after the rapture. Is it possible the completion of world evangelization could occur in fifty years?

The Olivet Discourse implies the rebuilding of the Jewish sacrificial Temple in which the abomination of desolation occurs as happening close to the same time as the completion of world evangelization. If the estimates by that Bible translation organization for the possible completion of translating the Scriptures into all remaining unreached people group languages in fifty years is realistic, this would suggest the possible completion of world evangelization within fifty years. Could fifty years be an approximate completion time for world evangelization? Does the Olivet Discourse imply the last major sign before the rapture, the abomination of desolation, is going to be fulfilled at approximately the same time as the completion of world evangelization? The conclusion is hard to avoid.

The estimate of fifty years from that Bible translation group was a human estimate lacking divine revelation. Murphy's law states that anything which can go wrong, will go wrong. On the other hand, it is possible that new circumstances, possibly even new technology, could develop which would facilitate and speed up the translation of the Scriptures to unreached people groups. In the meantime, it is appropriate to do what Jesus exhorted every true follower to do, "watch."

After the Apostles and disciples entered the fourth phase of Messiah's global evangelism strategy, as detailed in Acts chapter one, the possibility of a rapture at that time still seems highly improbable from Messiah's list of signs in the Olivet Discourse. Could those Apostles and disciples to whom Jesus predicted success in all four phases of His evangelism strategy realistically reach the majority of every people group on earth? Could the Apostles and disciples evangelize North and South America in their day? Could they have reached Australia or Hawaii? Could they, or their immediate converts, have evangelized every language people group of earth in their day? It seems highly improbable.

Jesus had predicted in the Olivet Discourse that completed world evangelization would occur before the end of the age, if not also before the rapture. Even after some of the Apostles and disciples reached and participated in all four phases of the evangelism strategy outlined to them by Jesus in Acts chapter one, it would appear that the possibility of an *unconditional, any-moment rapture requiring no intervening events* for those disciples and Apostles was still not a realistic possibility, at least from the perspective of the Olivet Discourse and its implied completion of world evangelization. In addition, the Scriptures seem to indicate there existed other prophecies and events which had to be fulfilled before the rapture could occur.

In the next chapter let us examine how the gift of prophecy relates to the idea of an *any-moment rapture requiring no intervening events.*

Chapter Ten

The Gift of Prophecy

> 6   And when Saul enquired of the LORD, the LORD answered him not, neither by dreams, nor by Urim, nor by prophets.

> 7   Then said Saul unto his servants, Seek me a woman that hath a familiar spirit, that I may go to her, and enquire of her. And his servants said to him, Behold, there is a woman that hath a familiar spirit at Endor.

> 8   And Saul disguised himself, and put on other raiment, and he went, and two men with him, and they came to the woman by night: and he said, I pray thee, divine unto me by the familiar spirit, and bring me him up, whom I shall name unto thee.

> 9   And the woman said unto him, Behold, thou knowest what Saul hath done, how he hath cut off those that have familiar spirits, and the wizards, out of the land: wherefore then layest thou a snare for my life, to cause me to die?

> 10   And Saul sware to her by the LORD, saying, As the LORD liveth, there shall no punishment happen to thee for this thing.

11 Then said the woman, Whom shall I bring up unto thee? And he said, Bring me up Samuel.

12 And when the woman saw Samuel, she cried with a loud voice: and the woman spake to Saul, saying, Why hast thou deceived me? for thou art Saul.

13 And the king said unto her, Be not afraid: for what sawest thou? And the woman said unto Saul, I saw gods ascending out of the earth.

14 And he said unto her, What form is he of? And she said, An old man cometh up; and he is covered with a mantle. And Saul perceived that it was Samuel, and he stooped with his face to the ground, and bowed himself.

15 And Samuel said to Saul, Why hast thou disquieted me, to bring me up? And Saul answered, I am sore distressed; for the Philistines make war against me, and God is departed from me, and answereth me no more, neither by prophets, nor by dreams: therefore I have called thee, that thou mayest make known unto me what I shall do.

16 Then said Samuel, Wherefore then dost thou ask of me, seeing the LORD is departed from thee, and is become thine enemy?

17 And the LORD hath done to him, as he spake by me: for the LORD hath rent the kingdom out of thine hand, and given it to thy neighbour, even to David:

18 Because thou obeyedst not the voice of the LORD, nor executedst his fierce wrath upon Amalek, therefore hath the LORD done this thing unto thee this day.

19 Moreover the LORD will also deliver Israel with thee into the hand of the Philistines: and to morrow shalt thou and thy sons be with me: the LORD also shall deliver the host of Israel into the hand of the Philistines. (I Samuel 28:6-19)

King Saul was Israel's first human king. Saul, at the start of his career, appeared to be a winner. He had apparent humility, and he was greatly honored in being chosen by the Lord to be Israel's first king. Saul seemed to be a man of success. King Saul became a loser. He disobeyed some of the Lord's instructions and was rejected as a source for Israel's future royalty. He was also rejected as the ancestor for the Messiah.

The above Scriptures indicate that when Saul sought the Lord for direction, the Lord refused to answer him. Since Saul couldn't get information about the future from the Lord, he resorted to forbidden methods in order to learn about the future, specifically through contact with a woman who was a medium for a demon.

Who wouldn't like to get some special inside information about the future? Astrological columns in newspapers eloquently attest to man's fascination with the future. People spend large sums of money to have horoscopes cast for them. The images of a woman with a crystal ball are familiar to most of us. People not only seek information about the future from the stars, they seek information about the future from palm readers, tarot cards, and Ouija boards.

Can these methods really obtain information about the future? Even if they can, does the Bible legitimize these types of procedures?

King Saul disguised himself to obtain the services of the woman who served as a medium for a demon. Saul himself had executed these mediums who offered their services, because such practices had been forbidden by the Lord. The Lord Himself had set the punishment for a medium as death. Saul was not ignorant of this rule, for he had carried out the capital punishment for offenders of this law as the witch at Endor herself indicated.

The Scriptures forbade Israel from participating in any of the cultic or occultic methods used to obtain information about the future. Any occultic or cultic method of obtaining information about the future was dishonoring to God, for it was done without God. People may not realize how serious these occult and cultic practices for obtaining knowledge about the future are to the Lord, but they should be aware that under the Old Covenant Scriptures, the penalty was execution. These methods are spiritually dangerous and often involve the operation of demons. These forbidden practices can lead the people who use them into self-destruction and spiritual danger. People engaging in forbidden methods to obtain knowledge about the future are at great risk concerning their eternal souls.

In this chapter we will examine how the gift of prophecy relates to the idea of an any-moment rapture requiring no intervening events.

Future Knowledge
Today, weathermen forecast climatic conditions for specific geographical areas based on a combination of factors that include weather patterns, air currents, temperature, and even historical data. Weathermen base their predictions on

known laws of science. They do not use occultic or spiritual methods for prediction, rather, they use scientific methodology. There is nothing wrong with scientific methods of forecasting. Jesus referred to weather prediction techniques which were quite successful in His lifetime, without any condemnation of those techniques.

> *54 And he said also to the people, When ye see a cloud rise out of the west, straightway ye say, There cometh a shower; and so it is.*
>
> *55 And when ye see the south wind blow, ye say, There will be heat; and it cometh to pass.*
>
> *56 Ye hypocrites, ye can discern the face of the sky and of the earth; but how is it that ye do not discern this time? (Luke 12:54-56)*

What Jesus condemned was the fact that some of the religious people had developed their skills of weather prediction with great accuracy, yet when it came to discerning spiritual signs, they were blind.

There have been methods used for predicting the future by people throughout history which were quite ungodly.

> *26 ... neither shall ye use enchantment, nor observe times. (Leviticus 19:26)*

The former [enchantment] refers to divination by serpents—one of the earliest forms of enchantment, and the other means the observation, lit., 'of clouds', as a study of the appearance and motion of clouds was a common way of foretelling good or bad fortune. Such absurd but deep-rooted superstitions often put a stop

to the prosecution of serious and important transactions, but they were forbidden especially as implying a want of faith in the being, or of reliance on the providence of God.
(JFB, n.d., p. 37)

9 When thou art come into the land which the LORD thy God giveth thee, thou shalt not learn to do after the abominations of those nations.

10 There shall not be found among you any one that maketh his son or his daughter to pass through the fire, or that useth divination, or an observer of times, or an enchanter, or a witch.

11 Or a charmer, or a consulter with familiar spirits, or a wizard, or a necromancer.

12 For all that do these things are an abomination unto the LORD: and because of these abominations the LORD thy God doth drive them out from before thee. (Deuteronomy 18:9-12)

What do some of these names for obtaining knowledge about the future mean?

> 'An observer of times.' ... 'one who augurs what is to happen;'... The word... is part of a verb which signifies to cover, to use covert arts, to practise sorcery;... 'one who divines by inspection–an augur.' 'An enchanter;' one who practises magic, or divines by signs... 'A witch'... probably one who pretended to cure diseases, or procure some desired result... The English word 'witch' is now restricted to the female practiser of unlawful arts; formerly it was applied to males as well,

if not chiefly... 'A charmer'... a dealer in spells, one who by means of spells or charms pretends to achieve some desired result. .. 'A consulter with familiar spirits.'... one who asks or inquires of ... [a] divining spirit. This spirit was supposed to be in the person of the conjurer, and to be able to reveal to him what was secret or hidden in the future... 'A necromancer;' one who professed to call up the dead, and from them to learn the secrets of futurity...
(Spence & Excell, n.d., Deuteronomy 18)

These practices were forbidden because they were intimately linked to idolatry. Seeking to obtain knowledge about the future through cultic or occultic methods is idolatry. Idolatry includes seeking information about the future from spiritual methods other than those involving or approved by the Lord.

Spiritual Future Knowledge
The Bible reveals that spiritual methods for obtaining knowledge about the future have been available throughout history. Sometimes this knowledge of the future was obtained by direct communion with God, as in the examples of Adam, Abraham, and Moses. This technique for gaining information about the future was a by-product of fellowship with the Lord. The Lord disclosed the future to them as a result of their communion with Him. That knowledge about the future strengthened and encouraged these people of faith on many occasions. Let us examine some Old Covenant examples of predictions and prophecies about the future which came through the Lord. This will help to prepare us later as we examine how the gift of prophecy relates to the rapture.

Adam received knowledge about the future through his

fellowship with the Lord.

> *16 And the LORD God commanded the man, saying, Of every tree of the garden thou mayest freely eat:*
>
> *17 But of the tree of the knowledge of good and evil, thou shalt not eat of it: for in the day that thou eatest thereof thou shalt surely die. (Genesis 2:16-17)*

Here a specific prediction about the future is made. The inference is that eating of any tree, except the tree of the knowledge of good and evil, would result in a continuing existence and a prolonged state of well-being for Adam. Eating of the forbidden tree would have detrimental consequences. We know that on the day Adam did eat of the forbidden tree, he did immediately experience spiritual death, which is alienation from the Lord. That spiritual death was later followed by physical death. So Adam did die on the day he ate of the tree, but it was a spiritual death which was later consummated with physical death.

Adam had the choice of either doubting the Lord's word, or of walking by faith and obeying the Lord's instructions. Adam lived by faith for some time, before he ended up disobeying the Lord. The following quote examines why death took place in the life of Adam. Was it because the tree had poisonous fruit, or was something else involved?

> The death that was to follow on transgression was to spring from the eating, and not from the fruit; from the sinful act, and not from the creature [plant], which in itself was good. The prohibition laid on Adam was for the time being a summary of the Divine law. Hence the tree was a sign and symbol of what that law required. And in this, doubtless, lies the explanation of its name.

It was a concrete representation of that fundamental distinction between right and wrong, duty and sin, which lies at the basis of all responsibility... Through its penalty it likewise indicated both the good which would be reaped by obedience and the evil which would follow on transgression."
(Spence & Exell, n.d., Gen. 2:16-17)

Abraham was a man who evidently communed with the Lord. He was from a land of idolatry, Ur of the Chaldees, which is in the territory better known as Babylon. In this land of idolatry and sin, Abram evidently communed with the real Lord. We are not informed as to how this recognition of the real God came about, but he heard from the Lord and was one day given instructions while in fellowship with God.

> *1 Now the LORD had said unto Abram, Get thee out of thy country, and from thy kindred, and from thy father's house, unto a land that I will shew thee:*
>
> *2 And I will make of thee a great nation, and I will bless thee, and make thy name great; and thou shalt be a blessing:*
>
> *3 And I will bless them that bless thee, and curse him that curseth thee: and in thee shall all families of the earth be blessed.*
>
> *4 So Abram departed, as the LORD had spoken unto him; and Lot went with him: and Abram was seventy and five years old when he departed out of Haran. (Genesis 12:1-4)*

Abraham had the opportunity to believe the Lord's predictions about the future, or he had the opportunity to doubt and to disbelieve. We are told that Abraham believed the Lord and walked by faith: he left Ur and headed toward the land which he had been told about from the Lord. He evidently believed these predictions about the future which the Lord had made to him. Abraham's trust in the Lord's predictions about the future resulted in his leaving Ur. When he arrived in Canaan, the Lord rewarded Abraham's faith with a special covenant. That covenant included some of the very same predictions made by the Lord to Abram before he left Ur.

1   And when Abram was ninety years old and nine, the LORD appeared to Abram, and said unto him, I am the Almighty God; walk before me, and be thou perfect.

2   And I will make my covenant between me and thee, and will multiply thee exceedingly.

3   And Abram fell on his face: and God talked with him, saying,

4   As for me, behold, my covenant is with thee, and thou shalt be a father of many nations.

5   Neither shall thy name any more be called Abram, but thy name shall be Abraham; for a father of many nations have I made thee.

6   And I will make thee exceeding fruitful, and I will make nations of thee, and kings shall come out of thee.

7   And I will establish my covenant between me and thee and thy seed after thee in their generations for an everlasting covenant, to be a God unto thee, and to thy seed after thee.

8   And I will give unto thee, and to thy seed after thee, the land wherein thou art a stranger, all the land of Canaan, for an everlasting possession; and I will be their God. (Genesis 17:1-8)

The Lord's predictions to Abraham included prophecies about the future in several areas. The predictions became part of a covenant with Abraham, because he had believed the Lord. The predictions which lured Abraham to Canaan were the promise of offspring, the prediction of a reputation, and a prediction of a condition (he would be a blessing and be blessed).

Kings did come from Abraham as the covenant here predicted. The greatest King of all kings came from Abraham: the Messiah.

There were other methods by which the Lord communicated to people knowledge of the future. In the Jewish Scriptures we see a prophetic gift at work in some of the patriarchs. Isaac, the son of Abraham, made amazing predictions about the future which were fulfilled by the Lord. These predictions were made in the form of blessings upon his sons, Jacob and Esau.

26   And his father Isaac said unto him, Come near now, and kiss me, my son.

27   And he came near, and kissed him: and he smelled the smell of his raiment, and blessed him, and said, See, the smell of my son is as the smell of a field which the LORD hath blessed:

28 Therefore God give thee of the dew of heaven, and the fatness of the earth, and plenty of corn and wine:

29 Let people serve thee, and nations bow down to thee: be lord over thy brethren, and let thy mother's sons bow down to thee: cursed be every one that curseth thee, and blessed be he that blesseth thee. (Genesis 27:26-29)

What is noteworthy is the fact that this prophetic announcement about the future was made by a blind man who thought he was blessing a totally different son. Jacob had deceived his father Isaac, pretending to be Esau. God was not deceived by the affair, so Isaac's prophetic anointing accurately predicted the future despite Isaac's ignorance as to the identity of the son receiving this blessing. This is revealed later by Isaac.

32 And Isaac his father said unto him, Who art thou? And he said, I am thy son, thy firstborn Esau.

33 And Isaac trembled very exceedingly, and said, Who? where is he that hath taken venison, and brought it me, and I have eaten of all before thou camest, and have blessed him? yea, and he shall be blessed. (Genesis 27:32-33)

Why did Isaac tremble so violently? He had evidently come to realize the deception, but he also recognized that the divine power which had anointed him to pronounce the blessing was from the Lord. This is why Isaac said that the person whom he had just blessed was going to be blessed. Isaac evidently felt the presence and supernatural anointing accompanying his pronouncement of the blessing upon Jacob the deceiver. This is evidently what terrified him, or caused him to quake so violently. He recognized that accompanying the blessing upon Jacob was a manifestation of the divine power and presence of the Lord. There was no doubt in Isaac's mind that the one whom he had blessed would receive the Lord's supernatural blessing because of the accompanying manifested divine presence as he uttered the blessing on that occasion.

Esau was horrified by his father Isaac's recognition that the blessing prophesied to Jacob could not be retracted. Isaac evidently realized that the blessing on Jacob had been sealed by the manifestation of the divine presence. So Esau cried out in spiritual agony, desiring a similar divine manifestation of prophetic blessing upon himself as well.

> 'and said unto his father, Bless me, even me also, O my father.' A proof of Esau's blind incredulity in imagining it to be within his father's power to impart benedictions promiscuously without and beyond the Divine sanction (Calvin)...
> (Spence & Exell, n.d., Gen. 27:34)

A similar anointing came upon Isaac as he blessed Esau.

> *37 And Isaac answered and said unto Esau, Behold, I have made him thy lord, and all his brethren have I given to him for servants; and with corn and wine have I sustained him: and what shall I do now unto thee, my son?*
>
> *38 And Esau said unto his father, Hast thou but one blessing, my father? bless me, even me also, O my father. And Esau lifted up his voice, and wept.*
>
> *39 And Isaac his father answered and said unto him,*

> *Behold, thy dwelling shall be the fatness of the earth, and of the dew of heaven from above;*
>
> *40 And by thy sword shalt thou live, and shalt serve thy brother; and it shall come to pass when thou shalt have the dominion, that thou shalt break his yoke from off thy neck. (Genesis 27:37-40)*

These predictions have been historically fulfilled, not by the persons Jacob and Esau, but by the nations which came from them. These prophecies were about two nations which would come from these two sons.

This same prophetic ability was later manifested by Jacob when he blessed his twelve sons before he died. A prophetic anointing supernaturally accompanied the pronouncements which came from his lips as he predicted the future of the tribes descending from each son. The pronouncement upon Judah is particularly worthwhile of examination, because it predicted the first coming of the Messiah. The Messianic nature of this prophetic declaration on Judah is recognized widely by both Jews and Christians.

> *9 Judah is a lion's whelp: from the prey, my son, thou art gone up: he stooped down, he couched as a lion, and as an old lion; who shall rouse him up?*
>
> *10 The sceptre shall not depart from Judah, nor a lawgiver from between his feet, until Shiloh come; and unto him shall the gathering of the people be. (Genesis 49:9-10)*

From here we find the origin for the imagery concerning the Messiah as "the lion of the tribe of Judah." These verses use the term "Shiloh" as a reference to the Messiah. That the term "Shiloh" is a reference to the Messiah and means "Sent" are indicated by a Latin translation and by a Jewish rabbi.

> ... the Vulgate, rendering... 'he that is sent,' and also... a rabbinical comment on Deut. xxii. 7: 'If you keep this precept, you hasten the coming of the Messiah who is called SENT.'"
> (Clarke, n.d., Gen. 49:10)

The verse which predicts the sceptre as not departing from Judah until the Messiah comes has been understood to indicate the following message.

> Judah shall continue a distinct tribe till the Messiah shall come; and it did so; and after his coming it was confounded with the others, so that all distinction has been ever since lost.
> (Clarke, n.d., Gen. 49:10)

There was another man who communed with God. Like Abraham, his fellowship with the Lord was on a personal level. Moses was a man who through communion with the Lord became aware of prophecies about the future. He also believed the Lord's predictions, and as a result, did great wonders which the Lord had prophesied.

> 2   And the LORD said unto him, What is that in thine hand? And he said, A rod.
>
> 3   And he said, Cast it on the ground. And he cast it on the ground, and it became a serpent; and Moses fled from before it.

4 And the LORD said unto Moses, Put forth thine hand, and take it by the tail. And he put forth his hand, and caught it, and it became a rod in his hand:

5 That they may believe that the LORD God of their fathers, the God of Abraham, the God of Isaac, and the God of Jacob, hath appeared unto thee.

6 And the LORD said furthermore unto him, Put now thine hand into thy bosom. And he put his hand into his bosom: and when he took it out, behold, his hand was leprous as snow.

7 And he said, Put thine hand into thy bosom again. And he put his hand into his bosom again; and plucked it out of his bosom, and, behold, it was turned again as his other flesh.

8 And it shall come to pass, if they will not believe thee, neither hearken to the voice of the first sign, that they will believe the voice of the latter sign.

9 And it shall come to pass, if they will not believe also these two signs, neither hearken unto thy voice, that thou shalt take of the water of the river, and pour it upon the dry land: and the water which thou takest out of the river shall become blood upon the dry land. (Exodus 4:2-9)

Many Old Covenant prophets received prophetic insight into the future concerning Israel's future, or the future of nations to which they prophesied. Daniel, Isaiah, Jeremiah, Ezekiel, Joel, Amos, Obadiah, Jonah, Zechariah, and Malachi all contain examples of prophetic insights into the future through divine means.

New Covenant Glimpses of the Future

Let us examine some New Covenant examples of prophecies and predictions about the future. This will help to prepare us as we examine the relationship between the gift of prophecy and the rapture of living and resurrected believers by the Messiah.

Divinely given insights into the future continued even during the times of Jesus and the early church. Jesus began to do miracles when He was anointed by the Holy Spirit, and He also began to predict the future. The first miracle Jesus did took place after His water baptism. This is indicated by the gospel writer John.

> *11 This beginning of miracles did Jesus in Cana of Galilee, and manifested forth his glory; and his disciples believed on him. (John 2:11)*

The miracles and prophetic pronouncements by Jesus appear to have begun after His anointing with the Holy Spirit, which took place at His water baptism. Prior to the Holy Spirit descending upon Jesus at His water baptism, Jesus did not do any miracles, and He did not have the gift of prophecy. Jesus had come as a human being. He was God, but He took upon Himself human form. As a human being He had limited His use of the divine attributes, including omniscience (all knowingness), so that He had to learn and develop as did other human beings. The anointing by the Holy Spirit which Jesus received at His water baptism was a prototype of the power the disciples and Apostles received on the day of Pentecost. The charismatic gifts which are dispensed to the church by the Holy Spirit were dispensed to Jesus by the Holy Spirit as well. After His water baptism, Jesus began to operate in the supernatural spiritual gifts which are available to all believers today.

13 Then cometh Jesus from Galilee to Jordan unto John, to be baptized of him.

14 But John forbad him, saying, I have need to be baptized of thee, and comest thou to me?

15 And Jesus answering said unto him, Suffer it to be so now: for thus it becometh us to fulfil all righteousness. Then he suffered him.

16 And Jesus, when he was baptized, went up straightway out of the water: and, lo, the heavens were opened unto him, and he saw the Spirit of God descending like a dove, and lighting upon him:

17 And lo a voice from heaven, saying, This is my beloved Son, in whom I am well pleased.

1 Then was Jesus led up of the spirit into the wilderness to be tempted of the devil. (Matthew 3:13-4:1)

It was after this event, that the supernatural miracles and ministry of Jesus began.

After the enduement of power was received by the disciples and Apostles on the day of Pentecost, they began to operate with the same supernatural gifts which Jesus had used after the Holy Spirit had come upon Him. There were instances of these supernatural miracles in the ministries of disciples before the day of Pentecost, but the day of Pentecost ushered in the era of power enduement for all believers through the charismatic gifts of the Holy Spirit.

Jesus had commanded His disciples to wait in Jerusalem for the promise of the Father. That promise was revealed by Jesus to be an enduement of power. The Father's promise would enable the disciples and apostles to use the same supernatural gifts Jesus had exercised. That power enduement came on the day of Pentecost.

> *1 And when the day of Pentecost was fully come, they were all with one accord in one place.*
>
> *2 And suddenly there came a sound from heaven as of a rushing mighty wind, and it filled all the house where they were sitting.*
>
> *3 And there appeared unto them cloven tongues like as of fire, and it sat upon each of them.*
>
> *4 And they were all filled with the Holy Ghost, and began to speak with other tongues, as the Spirit gave them utterance. (Acts 2:1-4)*

Not only did the apostles exercise these supernatural Holy Spirit gifts, these charismatic gifts were used by average believers as well. The promise of the Father was a reference to Joel's prophecy, as Peter explained on the day of Pentecost.

> *28 And it shall come to pass afterward, that I will pour out my spirit upon all flesh; and your sons and your daughters shall prophesy, your old men shall dream dreams, your young men shall see visions:*
>
> *29 And also upon the servants and upon the hand maids in those days will I pour out my spirit.*
>
> *30 And I will shew wonders in the heavens and in the earth, blood, and fire, and pillars of smoke.*

31 The sun shall be turned into darkness, and the moon into blood, before the great and terrible day of the LORD come.

32 And it shall come to pass, that whosoever shall call on the name of the LORD shall be delivered: for in mount Zion and in Jerusalem shall be deliverance, as the LORD hath said, and in the remnant whom the LORD shall call. (Joel 2:28-32)

This prophecy specifically predicts that God's Spirit would be poured out upon all ethnic groups on earth, which could only happen through world evangelism. Since the Apostles and disciples in New Covenant times did not evangelize all ethnic groups on earth, that prophecy by Joel could not be completely fulfilled until all ethnic groups were evangelized. Joel's prophecy indicates supernatural charismatic gifts did not stop after the Bible was written, but they are continuing down to our present day and age, so that all flesh, all ethnic groups, can experience sons and daughters prophesying. The reference to the moon as blood and the sun's darkness were apparently alluded to by Jesus in the Olivet Discourse. These celestial signs indicate Joel's prophecy extends down to the seven years of turmoil in the book of Revelation which compose the seven last years of the Gentile era.

The gift of prophecy is one of the charismatic gifts which the Holy Spirit dispenses to believers today. The gift of prophecy was exercised in Biblical times, and examples of the use of that gift are found in the New Covenant Scriptures. Particular examples of the charismatic gift of prophecy are found in the life of an early church believer named Agabus. It should be recognized that the charismatic gift of prophecy is not a gift which dispenses divine revelation equivalent to the Scriptures. The Bible indicates that the New Covenant gift of prophecy is inferior to the Scriptures in authority. The gift of prophecy can operate to supernaturally foretell the future by divine power. This foretelling manifestation of the gift has clear Biblical examples.

> *27 And in these days came prophets from Jerusalem unto Antioch.*
>
> *28 And there stood up one of them named Agabus, and signified by the Spirit that there should be great dearth throughout all the world: which came to pass in the days of Claudius Caesar.*
>
> *29 Then the disciples, every man according to his ability, determined to send relief unto the brethren which dwelt in Judaea: (Acts 11:27-29)*

In this example of the foretelling manifestation of the gift of prophecy, Agabus supernaturally predicted a famine. This supernatural foreknowledge notified believers in Antioch that a famine was about to come upon the world.

> *10 And as we tarried there many days, there came down from Judaea a certain prophet, named Agabus.*
>
> *11 And when he was come unto us, he took Paul's girdle, and bound his own hands and feet, and said, Thus saith the Holy Ghost, So shall the Jews at Jerusalem bind the man that owneth this girdle, and shall deliver him into the hands of the Gentiles.*
>
> *12 And when we heard these things, both we, and they of that place, besought him not to go up to Jerusalem. (Acts 21:10-12)*

In the above passage of Scripture, Agabus exercised the gift of prophecy again in a foretelling manifestation. He predicted that the Apostle Paul would be taken prisoner in Jerusalem.

The gift of prophecy, as manifested in the above Biblical examples, enabled the man Agabus to supernaturally foretell specific events about the future.

The charismatic gifts of the Holy Spirit have been manifested throughout church history, and are widely manifested throughout the world today among Pentecostal and Charismatic groups of believers. The supernatural miracle ministries of two prophets predicted in the book of Revelation, chapter eleven, will manifest the charismatic gift of prophecy, as well as the gift of miracles. While some believe those two men are Moses and Elijah, or Enoch and Elijah returned to earth, it seems more likely, in view of the Elijah prophecies fulfilled by John the Baptist, that they will be two Charismatic or Pentecostal men living at that time of future history.

Some groups of believers today who claim the Bible as their authority will deny that the present day worldwide manifestations of spiritual gifts among Charismatic and Pentecostal believers are a continuation of the enduement of power which began on the day of Pentecost as recorded in Acts chapter two. The number of "Christian" groups denying the present day manifestation of the Biblical charismatic gifts appears to be shrinking. These Biblical spiritual gifts are manifested throughout the world on such a tremendous scale, that there are few Christians today who are unaware of these claimed spiritual gifts in well-known Christian ministries.

While some groups claiming the Bible as their source of authority state that some or all of the charismatic spiritual gifts of I Corinthians 12 have ended in New Testament times, there is division among those groups over which of the gifts have ceased, and there is division over the Biblical passages which are used for claiming those gifts have stopped.

Have Spiritual Gifts Stopped?
While the purpose of this book is not to discuss whether the Charismatic or Pentecostal gifts are today valid manifestations of the Biblically defined gifts of I Corinthians 12, the subject has relevance to the study of Biblical prophecy. In the next section the reason will be revealed as to why the manifestation of the gift of prophecy has real value to the study of Biblical prophecy concerning the rapture.

The groups are shrinking which claim that the I Corinthians 12 spiritual gifts are not valid today. Some claim those gifts died out in the times of the apostles. Some claim those gifts stopped altogether when the Bible was finally completed. While some groups claim these miraculous sign gifts have stopped, almost everyone is aware of present day Christian ministries claiming the supernatural I Corinthians 12 supernatural gifts are today active and manifested in their own ministries. Groups claiming those Biblical sign gifts have ceased either deny contemporary manifestations are valid, or more dangerously to their own spiritual well-being, claim divine manifestations are of satanic origin.

Pentecostals and Charismatics claim the I Corinthians 12 spiritual gifts are continuing today, and that those charismatic gifts are manifested worldwide. The largest churches in the world today are Pentecostal churches.

The controversy is over when the I Corinthians 12 miraculous gifts stop, because almost everyone universally agrees the Bible indicates at least some, if not all, these miraculous gifts will stop at some time. The evidence for this can be found in the following Scriptures.

> *8 Charity never faileth: but whether there be prophecies, they shall fail; whether there be tongues, they shall cease; whether there be knowledge, it shall vanish away.*
>
> *9 For we know in part, and we prophesy in part.*
>
> *10 But when that which is perfect is come, then that which is in part shall be done away.*
>
> *11 When I was a child, I spake as a child, I understood as a child, I thought as a child: but when I became a man, I put away childish things.*
>
> *12 For now we see through a glass, darkly; but then face to face: now I know in part; but then shall I know even as also I am known. (I Corinthians 13:8-12)*

These Scriptures indicate that at least some, if not all, the charismatic spiritual gifts will one day stop. The charismatic gifts will stop when the "perfect" (13:10) is come. What is the "perfect?"

Some claim that the "perfect" is the canon of Scripture (the Bible). These people claim that when the Bible was completed, the I Corinthians 12 miraculous gifts stopped. There is a problem with this interpretation, because Biblical evidence indicates that prophecy will continue even into the millenium. Prophecy is also manifested in the ministries of the two men who smite the earth with plagues as often as they will during the last seven years of the Gentile era which are found in the book of Revelation.

> *3 And I will give power unto my two witnesses, and they shall prophesy a thousand two hundred and threescore days, clothed in sackcloth.*
>
> *4 These are the two olive trees, and the two candlesticks standing before the God of the earth.*
>
> *5 And if any man will hurt them, fire proceedeth out of their mouth, and devoureth their enemies: and if any man will hurt them, he must in this manner be killed.*
>
> *6 These have power to shut heaven, that it rain not in the days of their prophecy: and have power over waters to turn them to blood, and to smite the earth with all plagues, as often as they will.*
> *(Revelation 11:3-6)*

These two witnesses evidently are men who exist at that time and who manifest the I Corinthians 12 spiritual gifts of prophecy and miracles. There are also Scriptures suggesting that during the future thousand year reign of the Messiah, the gift of prophecy will continue.

In view of the above Biblical data, it seems highly unlikely that the "perfect" should be defined as "the completed canon of Scripture." It is not the completion of the Bible which causes spiritual gifts to stop. It is more likely that the "perfect" is the eternal state, when believers "will know even as they are known."

> *12 For now we see through a glass, darkly; but then*

face to face: now I know in part; but then shall I know even as also I am known. (I Corinthians 13:12)

When believers have entered the eternal state subsequent to the millenial kingdom, they will be dwelling with the Lord face to face. The Lord will be in their midst. Why would they need the gift of prophecy then?

1 And I saw a new heaven and a new earth: for the first heaven and the first earth were passed away; and there was no more sea.

2 And I John saw the holy city, new Jerusalem, coming down from God out of heaven, prepared as a bride adorned for her husband.

3 And I heard a great voice out of heaven saying, Behold, the tabernacle of God is with men, and he will dwell with them, and they shall be his people, and God himself shall be with them, and be their God.

4 And God shall wipe away all tears from their eyes; and there shall be no more death, neither sorrow, nor crying, neither shall there be any more pain: for the former things are passed away.

22 And I saw no temple therein: for the Lord God Almighty and the Lamb are the temple of it.

23 And the city had no need of the sun, neither of the moon, to shine in it: for the glory of God did lighten it, and the Lamb is the light thereof.
(Revelation 21:1-4, 22-23)

The above Scriptures appear to define what the "perfect" is

which causes the charismatic spiritual gifts to stop. It is the personal presence of God Himself in the New Heavens and New Earth which constitutes the "perfect." At that time we see the Lord "face to face."

> *12 For now we see through a glass, darkly; but then face to face: now I know in part; but then shall I know even as also I am known. (I Corinthians 13:12)*

The I Corinthians 12 gift of prophecy will stop, according to the Scriptures, when the perfect is come. When we no longer "know in part," then prophecy will stop. When we see the Lord "face to face" in eternity, in the New Jerusalem, we won't need prophecy on earth any more. The gift of prophecy of I Corinthians 12 will no longer be necessary in that day.

Why this concern over the I Corinthians 12 spiritual gifts which have often been called the Charismatic or Pentecostal gifts? Why the concern over when the spiritual gifts of I Corinthians 12 stop? The reason for this concern will be made evident in the next section.

The Gift of Prophecy and Foretelling

The I Corinthians 12 gift of prophecy was manifested in the ministry of Agabus who foretold a famine (Acts 11:27-29), and who also foretold that Paul would be taken prisoner (Acts 21:10-12). The I Corinthians 12 gift of prophecy was evidently manifested, according to the apostle Paul, in a number of cities.

> *22 And now, behold, I go bound in the spirit unto Jerusalem, not knowing the things that shall befall me there:*

23 Save that the Holy Ghost witnesseth in every city, saying that bonds and afflictions abide [await] me. (Acts 20:22-23)

It wasn't Agabus alone who foretold Paul's imprisonment at Jerusalem. Evidently, believers in every city Paul had passed through had manifested the gift of prophecy in a foretelling manner, and had predicted Paul's imprisonment at Jerusalem. Paul, according to the gift of prophecy, would be taken prisoner.

No matter how one interprets Biblical passages concerning the time at which the prophetic gifts will stop, there is one issue on which all believers should have little difficulty agreeing. In the New Covenant Scriptures we see a definite manifestation of the I Corinthians 12 gift of prophecy in the ministry of Agabus, and according to Paul, in other cities through other believers as well. The manifestation of the I Corinthians 12 charismatic gift of prophecy took as its form the foretelling of future events. These predicted events by Agabus included the prediction of a famine, and a prediction of Paul's imprisonment at Jerusalem. Since this gift of prophecy was a manifestation of the Holy Spirit's knowledge about the future through Agabus and others, was the rapture a possible event before the Apostle Paul was taken prisoner?

The answer is, "No." The rapture was not possible for Paul the Apostle, nor for any other believer, while Paul was on his way to Jerusalem, and while Paul had not yet been taken prisoner in the form predicted by the Holy Spirit. The issue is this, can an *any-moment, imminent rapture requiring no intervening events* be considered an accurate teaching when the Holy Spirit has made a prediction foretelling a future earthly event in the life of a believer? In other words, is an *imminent, any-moment rapture* accurate if the Holy Spirit predicted an event in a believer's life which must be fulfilled before the rapture? If something must be fulfilled before the rapture, how can the rapture happen at any-moment?

If the I Corinthians 12 gift of prophecy is manifested today, as in Biblical times, in a foretelling manner which predicts a future event in the life of a Christian, can the rapture be considered to be an "imminent, any-moment" possibility when there has been a prediction of an intervening event by the Holy Spirit? The any-moment rapture was a logical impossibility before Paul had been taken prisoner in Jerusalem in the manner Agabus and others had predicted through the Holy Spirit. What about today?

Is the I Corinthians 12 gift of prophecy operative today also? If it is, can a Holy Spirit predicted event in the life of a believer postpone the rapture?

It does appear from the Scriptures that the I Corinthians 12 gift of prophecy is in operation today. Since it is operative, the question becomes, "If the Holy Spirit predicts events today in believers lives, does that mean the rapture is postponed until those events are fulfilled?" Is an any-moment, sign-less rapture possible while a Holy Spirit predicted event in a believer's life is unfulfilled?

Some may be tempted to suggest that since we are to test prophecy (I John 4:1), we may not always be able to discern in advance whether an apparent manifestation of the gift of prophecy is legitimately from the Holy Spirit. If the Holy Spirit did not make the prediction about some future event in a person's life, and the prediction was really a counterfeit, this would mean the any-moment rapture would still be possible, because a specific manifestation might really be a false or phony prediction about the future. This evaluation amounts to the concept that since we cannot be certain about the source of the prophecy before it is fulfilled, the gift of prophecy in a predictive manner is still in harmony with an any-moment rapture.

In reply, it should be stated that if a prediction of some event which must take place before the rapture is phony, it does not affect the truth or error of an any-moment teaching. I will restate the problem in another manner to highlight the real issue: is a I Corinthians 12 charismatic gift of prophecy about a future event in a believer's life even theoretically possible if an *any-moment, imminent rapture requiring no intervening events* is a truly Biblical teaching? To simplify the issue, we can restate it a little differently. The gift of prophecy is often manifested to predict intervening events in a believer's life, events which must be fulfilled before the rapture. Are intervening events even theoretically possible if the teaching of an any-moment rapture is accurate?

If a person answers, "Yes," to this question, I believe they have failed to grasp the idea that a rapture which can happen at *any-moment, with no intervening events required*, is a teaching which theoretically negates even the possibility of a I Corinthians 12 manifestation of predictive prophecy concerning a future event in the life of a believer. Let us restate this more simply: the prediction of intervening events is not possible, if the teaching that the rapture can happen at any-moment is accurate. Or, to state it another way: predicting events which must occur before the rapture is not possible if the rapture can happen at any-moment.

The I Corinthians 12 gift of prophecy, as manifested by

Agabus, appears to be contradictory to an any-moment rapture purely on a theoretical level. The two concepts are mutually exclusive, and mutually contradictory. If the two concepts were mutually exclusive and mutually contradictory during New Covenant times, it certainly seems that the two concepts continue to be mutually contradictory even today. Either only one of these teachings is legitimate, or the Bible contradicts itself. Either the any-moment rapture requiring no intervening events is an accurate teaching, or the contemporary I Corinthians 12 gift of prophecy manifested in a predictive manner in the lives of believers today is an accurate teaching. These two concepts exclude each other, and are incompatible. Either one or the other is true, but they cannot both be true at the same time. The gift of prophecy, as Biblically illustrated in the life of Paul the Apostle, is often manifested through the prediction of specific events in believers' lives which must be fulfilled before the rapture can occur. The prediction of events which must occur before the rapture contradicts the idea that the rapture can happen at any-moment, and it contradicts the idea that no intervening events are required for the rapture to take place. The gift of prophecy often foretells intervening events. Intervening events indicate the rapture cannot occur at any-moment.

Now some may still try to appeal to the argument of ignorance. They may restate that since a believer might not be able to discern whether a prediction came from the Holy Spirit, an imminent rapture and the gift of prophecy are harmonious. In examining this appeal to ignorance, we can ask: is this a pragmatic approach, or is it an attempt to rationalize and harmonize mutually contradictory ideas? Let us examine this argument to see if it really is valid. Let us restate it a little differently to analyze it from a slightly different perspective. What the argument really is suggesting is the following: a believer may not know whether it was the Holy Spirit which predicted an event which must be fulfilled before the rapture, so it might be a phony prediction, so the prediction of events which must occur before the rapture is still harmonious with the idea that no events can prevent the rapture. To simplify that attempt to harmonize an imminent rapture and the gift of prophecy we can state it another way. The argument states: since we don't know if it was the Holy Spirit that predicted an event which must occur before the rapture, the teaching that the rapture can occur at any-moment is still correct. That appeal to ignorance really changes the whole subject. The real issue is not whether a prediction is a false prophecy, the real issue is, what if the prediction is from the Holy Spirit?

It must be admitted, the plea to ignorance is accurate when it states that a phony prediction does not determine whether the rapture can happen at any-moment or not. Still, that is not the issue of concern. The real issue is not whether a prediction is phony. The real issue is, what if the prediction is from the Lord? If the Holy Spirit predicts an event in a believer's life which precedes the rapture, how can the rapture happen at any-moment? Obviously, stated this way, the appeal to ignorance is a smoke screen. The appeal to ignorance really changes the subject. The issue is not whether the prediction is phony. The issue is this: what if the Holy Spirit foretells a future event which must take place before the rapture? The gift of prophecy is often manifested to predict intervening events in a believer's life, events which must be fulfilled before the rapture. Are intervening events even theoretically possible if the teaching of an any-moment rapture is accurate?

Charismatic and Pentecostal groups which claim the I

Corinthians 12 supernatural gift of prophecy is a legitimate gift for today, are shooting themselves in the foot, so to speak, by also endorsing an imminent rapture theory requiring no intervening events. The I Corinthians 12 gift of prophecy, when operated in the same predictive manner as Agabus displayed, contradicts the idea of an any-moment rapture. The gift of prophecy was manifested in the New Covenant Scriptures as a Holy Spirit gift which predicted events in believers' lives which had to occur before the rapture. The gift of prophecy often foretells intervening events. If intervening events can be foretold, it is not accurate to say the rapture can happen at any-moment. If intervening events can be foretold, how can one say that the rapture has no required intervening events? If the Holy Spirit foretold the intervening event, the intervening event is required to be fulfilled before the rapture can take place.

Mutually Contradictory

A detailed study of the modern, *any-moment rapture theory requiring no intervening events*, will reveal that the teaching is a peripheral development of pre-tribulationalism. As we have already seen in an earlier chapter, the imminency idea came into existence after the Millerite movement had met with great failure in its attempts to predict the time of the rapture. The doctrine of "imminency," as it came to be called, appears to have been transformed by a committee into an anti-date setting formula. The "imminency" teaching of an *any-moment rapture theory requiring no intervening events* was heavily influenced by the pre-tribulational rapture theory. What may not be well known is the fact that John Nelson Darby, the formulator of the pre-tribulational rapture theory, was staunchly opposed to the modern manifestation of the Pentecostal/Charismatic gifts (Lindsey, 1999, p. 132). It would appear that the "imminency" teaching, which developed out of an environment heavily influenced by Darby's ideas, manifests the same anti-Pentecostal and anti-Charismatic preferences Darby held.

It is strange that "imminency" is embraced by Pentecostal and Charismatic groups today. "Imminency" has contradictory concepts which oppose the manifestation of the gift of prophecy as practiced by Agabus. Agabus used the gift of prophecy in the New Covenant Scriptures to predict intervening events in believers' lives, that is, events which had to occur before the rapture. Other believers in numerous cities also predicted, through the power of the Holy Spirit, an intervening event in the life of the Apostle Paul. That intervening event was the prediction that Paul would be taken prisoner in Jerusalem before the rapture.

It is not too difficult to understand how some Pentecostal and Charismatic groups may have come to adopt the "imminency" teaching which contradicts modern manifestations of the gift of prophecy. The reason these groups adopted imminency is probably based on the fact that they never thought through the implications of the imminent rapture teaching as it related to the gift of prophecy. In addition, they may also have been eager to avoid the stain and stigma and problems associated with people who might attempt to set dates for the rapture as William Miller did. Date setters would have tarnished the reputations of denominations, churches, or organizations with their false teachings. Imminency is a potent anti-date setting formula. Imminency, when adopted and enforced, does effectively discourage date-setting for the rapture. Solving the erroneous tendency to set dates was a good move. Solving an erroneous tendency with a teaching which is itself erroneous, is an error. The "imminency" teaching solves one problem, but it overcorrects the error. Overcorrecting an error is itself an error.

It is important to recognize that the "imminent" rapture theory is at odds with the contemporary manifestation of the gift of prophecy. The two concepts are at odds with each other. It is possible for organizations to maintain mutually contradictory ideas for temporary periods of time, but Jesus stated, "And if a kingdom be divided against itself, that kingdom cannot stand." For organizations to maintain mutually contradictory ideas, there may be a perception on the part of honest examiners that the organization has actually mandated a requirement for intellectual suicide from its members.

I have chatted with Christian believers and leaders who taught, or were convinced about, the truth of an any-moment, imminent rapture. Several were notified that I didn't share their view, and that I considered the teaching un-biblical. Some have already died. They were not looking for the undertaker at the time I spoke with them about the issue. One said definitely that he was "looking for the upper-taker." I assume he was referring to the rapture.

For most believers throughout history, the rapture has not been a Biblically valid possibility. The time is getting closer for the rapture to become a valid possibility. As of the moment of this writing, there remain yet some unfulfilled signs which appear to require fulfillment for the rapture to become a valid option.

All believers should be watching for the Lord to come. We should be watching for the signs He specified. Those signs will help believers to be ready for that moment when the rapture trumpet will sound.

In the next chapter an attempt will be made to examine the rapture as it relates to events predicted by an angel.

Chapter Eleven

Angel of Hope

The Mediterranean Sea can at times be rather hazardous, especially when tempestuous winds blow upon it. Near the coasts these winds can become a danger to shipping. Waves in the Mediterranean can reach a height of twenty feet from trough to crest. A combination of wind and waves could quickly result in life threatening situations for sailors, especially for sailors who lived in the time of Paul the Apostle.

In the book of Acts, Luke recounts an incident out of the life of the Apostle Paul. While being transported by ship to Rome as a prisoner, Paul endured a harrowing experience on the Mediterranean, which came close to taking the lives of everyone on board the ship on which Paul was a passenger.

The Apostle had been taken prisoner at Jerusalem just as had been predicted by the Holy Spirit through the prophet Agabus.

> 10 And as we tarried there many days, there came down from Judaea a certain prophet, named Agabus.
>
> 11 And when he was come unto us, he took Paul's gir-

dle, and bound his own hands and feet, and said, Thus saith the Holy Ghost, So shall the Jews at Jerusalem bind the man that owneth this girdle, and shall deliver him into the hands of the Gentiles.

12 And when we heard these things, both we, and they of that place, besought him not to go up to Jerusalem. (Acts 21:10-12)

Numerous other Christians had also predicted through the power of the Holy Spirit that Paul would be taken prisoner in Jerusalem.

22 And now, behold, I go bound in the spirit unto Jerusalem, not knowing the things that shall befall me there:

23 Save that the Holy Ghost witnesseth in every city, saying that bonds and afflictions abide [await] me. (Acts 20:22-23)

Rather than agree to a confrontation with the religious leaders of Jerusalem, we understand that Paul had appealed to Caesar from the account in Acts.

9 But Festus, willing to do the Jews a pleasure, answered Paul, and said, Wilt thou go up to Jerusalem, there be judged of these things before me?

10 Then said Paul, I stand at Caesar's judgment seat, where I ought to be judged: to the Jews have I done no wrong, as thou very well knowest.

11 For if I be an offender, or have committed any thing worthy of death, I refuse not to die: but if there be none of these things whereof these accuse me, no man may deliver me unto them. I appeal unto Caesar.

12 Then Festus, when he had conferred with the council, answered, Hast thou appealed unto Caesar? unto Caesar shalt thou go. (Acts 25:9-12)

The appeal was perhaps a self-preservation maneuver, for Paul's life had already been threatened by a discovered plot to ambush and murder him. The nefarious plot was formed by a group of the Jews in Jerusalem. The attempt to get Paul to face charges in Jerusalem could possibly have been evaluated by the Apostle as a potentially disguised ploy to lure him back to that city in order to provide the band of self-appointed assassins an opportunity to take his life. The following Scriptures indicate the discovered plot against the Apostle.

12 And when it was day, certain of the Jews banded together, and bound themselves under a curse, saying that they would neither eat nor drink till they had killed Paul.

13 And they were more than forty which had made this conspiracy.

14 And they came to the chief priests and elders, and said, We have bound ourselves under a great curse, that we will eat nothing until we have slain Paul.

15 Now therefore ye with the council signify to the chief captain that he bring him down unto you to morrow, as though ye would enquire something more perfectly concerning him: and we, or ever he come near, are ready to kill him.

16 And when Paul's sister's son heard of their lying in wait, he went and entered into the castle, and told Paul. (Acts 23:12-16)

Festus may have simply been attempting to create a public demonstration to please Paul's accusers which he may have been unable to actually fulfill. It may be that Festus did not have the right to judge Paul in Jerusalem. The following comment indicates that the plan could have been to hand the Apostle over to the religious leaders in Jerusalem.

... he understood Festus to propose handing him over to the Sanhedrin for judgment... with a mere promise of protection from him. But from going to Jerusalem at all he was too well justified in shrinking for there assassination had been recently planned against him. (JFB, n.d., Acts 25:9-10)

Paul had been taken prisoner due to a public disturbance.

27 And when the seven days were almost ended, the Jews which were of Asia, when they saw him in the temple, stirred up all the people, and laid hands on him,

28 Crying out, Men of Israel, help: This is the man, that teacheth all men every where against the people, and the law, and this place: and further brought Greeks also into the temple, and hath polluted this holy place.

29 (For they had seen before with him in the city Trophimus an Ephesian, whom they supposed that Paul had brought into the temple.)

30 And all the city was moved, and the people ran

together: and they took Paul, and drew him out of the temple: and forthwith the doors were shut.

31 And as they went about to kill him, tidings came unto the chief captain of the band, that all Jerusalem was in an uproar.

32 Who immediately took soldiers and centurions, and ran down unto them: and when they saw the chief captain and the soldiers, they left beating of Paul.

33 Then the chief captain came near, and took him, and commanded him to be bound with two chains; and demanded who he was, and what he had done.

34 And some cried one thing, some another, among the multitude: and when he could not know the certainty for the tumult, he commanded him to be carried into the castle.

35 And when he came upon the stairs, so it was, that he was borne of the soldiers for the violence of the people.

36 For the multitude of the people followed after, crying, Away with him. (Acts 21:27-36)

Paul could have been set free, but he exercised his privileges as a Roman citizen and requested a review of his case by none other than the head of the Roman Empire, Caesar. Paul's request for an audience with Caesar was granted, and the Apostle was in the process of being escorted as a prisoner to Rome when a fierce wind destroyed the very ship in which he was sailing.

Paul had advised his Roman captors not to sail because of the lateness of the season. Against Paul's advice, the Romans and sailors set out to sea and ended up in a tremendous windstorm.

> *9 Now when much time was spent, and when sailing was now dangerous, because the fast was now already past, Paul admonished them,*
>
> *10 And said unto them, Sirs, I perceive that this voyage will be with hurt and much damage, not only of the lading and ship, but also of our lives.*
>
> *11 Nevertheless the centurion believed the master and the owner of the ship, more than those things which were spoken by Paul.*
>
> *12 And because the haven was not commodious to winter in, the more part advised to depart thence also, if by any means they might attain to Phenice, and there to winter; which is an haven of Crete, and lieth toward the south west and north west.*
>
> *13 And when the south wind blew softly, supposing that they had obtained their purpose, loosing thence, they sailed close by Crete.*
>
> *14 But not long after there arose against it a tempestuous wind, called Euroclydon.*
>
> *15 And when the ship was caught, and could not bear up into the wind, we let her drive. (Acts 27:9-11)*

For several days the ship was tossed about at sea so those aboard lost all hope of escaping the storm alive.

> *18 And we being exceedingly tossed with a tempest, the next day they lightened the ship;*
>
> *19 And the third day we cast out with our own hands the tackling of the ship.*
>
> *20 And when neither sun nor stars in many days appeared, and no small tempest lay on us, all hope that we should be saved was then taken away.*
> *(Acts 27:18-20)*

The following comment indicates some of the difficult conditions the passengers and crew were facing in this very distressing situation.

> 'neither sun nor stars appeared many ('several') days'–probably most of the fourteen days mentioned... This continued thickness of the atmosphere prevented their making the necessary observations of the heavenly bodies by day or by night; so that they could not tell where they were... 'Their exertions to subdue the leak had been unavailing; they could not tell which way to make for the nearest land, they must founder at sea. Their apprehensions, therefore, were not so much caused by the fury of the tempest, as by the state of the ship.' [SMITH.] From the inferiority of ancient to modern naval architecture, leaks were sprung much more easily, and the means of repairing them were fewer than now. Hence the far greater number of shipwrecks from this cause.
> (JFB, n.d., Acts 27:18-20)

It is at this point we see the astounding value of a spiritual relationship with God, as exhibited in the life of the Apostle to the Gentiles, Paul. For some time Paul fasted before God during the storm, resulting in a supernatural visitation to him. An angel of God appeared to Paul aboard the ship and assured him that he would survive the storm. The angel further revealed that all of the people aboard the ship would also survive as a result of Paul's fasting and prayer.

> *21 But after long abstinence Paul stood forth in the midst of them, and said, Sirs, ye should have hearkened unto me, and not have loosed from Crete, and to have gained this harm and loss.*
>
> *22 And now I exhort you to be of good cheer: for there shall be no loss of any man's life among you, but of the ship.*
>
> *23 For there stood by me this night the angel of God, whose I am, and whom I serve, (Acts 27:21-23)*

That same night Paul related the good news to all those aboard the ship. He began by berating his captors for not listening to his advice about the bad sailing weather. "I told you so," would be the present day equivalent to his introductory message of hope to those aboard the vessel.

"You should have listened to me when I told you not to sail from Crete," Paul scolded. "I do have some good news for you despite the present situation," the Apostle said encouraging them. "You are all going to survive this horrible storm, because the God I serve sent an angel to tell me that tonight."

This would be a remarkable example of God's concern for human beings even if this were all that had occurred in this incident preserved for us in the book of Acts. There is something else that happened during this trip which has long-range repercussions. We should be aware of this. Something that actually happened aboard the sailing vessel relates to the coming of the Messiah. Paul's spiritual experience of a visitation from an angel of God has real significance concerning the coming of the Messiah in a very practical way.

An Angel Speaks
The angel which supernaturally appeared to the Apostle Paul aboard ship during the terrible storm had a message for Paul that had significance not only for the Apostle, but it has significance for us today also.

The angel which appeared to Paul aboard the ship assured the Apostle that he would survive. What a comfort this angel must have been! The angel's message to Paul was one of assurance during a time of intense stress.

There is much more that the angel had to share with Paul. We should be aware of the other part of the angel's message to the Apostle. The angel's second half of the message is deeply moving when it comes to analyzing the Apostle's personal viewpoint on the coming of the Messiah. If we truly want to understand the Apostle Paul and his personal views on the rapture rescue of Messiah's believers from the earth, then we must listen to the angel's words.

> *23 For there stood by me this night the angel of God, whose I am, and whom I serve,*
>
> *24 Saying, Fear not, Paul; thou must be brought before Caesar: and, lo, God hath given thee all them that sail with thee.*

25 Wherefore, sirs, be of good cheer: for I believe God, that it shall be even as it was told me.

26 Howbeit we must be cast upon a certain island. (Acts 27:23-26)

Astounding words! More astounding is the significance of these words toward the practical question of the coming of the Messiah at the rapture rescue of His true followers. Let us examine specifically the angel's message of comfort concerning the preservation of the life of the Apostle Paul. The angel's specific words were, "... thou must be brought before Caesar..."

Did the Apostle have an Imminent Hope?

It has been widely circulated among religious circles that the Apostle Paul personally believed in and taught an "any-moment" coming of the Messiah at the rapture rescue of true believers from the earth. This teaching states that the Messiah could come for His followers at "any time," even in the next couple of seconds. That teaching has been described as being an event which is "imminent." It has been widely circulated and taught that Paul believed Jesus would come again at any time to rescue Paul and other genuine followers of Jesus from the earth, even at any second during the lifetime of the Apostle. If this were actually an accurate belief, what did the angel's words mean? The angel had spoken to the Apostle, and according to the writer Luke, the angel made the following statement, "Fear not, Paul; thou must be brought before Caesar... "

The angel informed Paul that he would personally survive the shipwreck. Even more than this, the angel gave Paul a brief summary of his personal future. According to the angel, not only would Paul survive the coming shipwreck, he and the others would also remain on earth to be cast upon an island. In addition, not only would he live and remain on earth to be cast upon an island, he would live and remain on earth long enough to one day personally stand before the leader of the Roman Empire.

In relating these predictions about the future to the Apostle Paul, the angel intimated some facts about Paul's relationship to the rapture which do not quite agree with the popular teaching that he was looking for an any-moment coming of the Messiah to snatch away believers from the earth. The angel's words indicate that contrary to the popular view that Paul was personally looking for an any-moment rapture which would cause genuine believers to rise into the clouds, the Apostle had been informed that he would remain on earth for an extended period of time, and that he would remain on this earth long enough to stand before the Emperor at Rome.

The angel's message to the Apostle was a sort of combination of good news and bad news. The good news was that Paul would live. The good news was that Paul would live to be cast upon an island. The good news was that Paul would stand before the throne of the Roman leader one day. The bad news was that in the meantime he should forget about an *any-moment rapture requiring no intervening events.* The angel's message had preempted the possibility of the rapture for at least a certain segment of the Apostle's lifetime.

As long as Paul had not had the opportunity to stand before Caesar, the Apostle could be quite confident that the Messiah's sudden coming in the clouds to snatch away living believers from earth would not be a possible event.

Was the angel who appeared to the Apostle a liar about the delayed rapture? Elsewhere, Paul has written about angels in some very scathing words.

> 8 But though we, or an angel from heaven, preach any other gospel unto you than that which we have preached unto you, let him be accursed.
> (Galatians 1:8)

In this verse from Galatians the Apostle specifically indicates that angels should not necessarily be believed. He indicates in that verse that there are occasions in which angels should be doubted. He particularly indicates that statements by angels are to be rejected if their statements do not agree with the previously preached gospel.

Why did Paul believe the angel which appeared to him? Maybe Paul should have been skeptical of the angel's message, or perhaps Paul was having some sort of hallucination. The Apostle had been fasting for a lengthy period of time. Maybe the devil was playing tricks on him. According to many Bible teachers, the Apostle believed in, and apparently taught, an *any-moment rapture which required no intervening events.* This angel claiming to be from God was predicting intervening events. Didn't the angel know any better than to contradict what Paul supposedly preached?

If the angel which appeared to the Jew from Tarsus aboard that vessel was contradicting what Paul supposedly preached *(an any-moment rapture requiring no intervening events),* why did the Apostle believe the angel which contradicted him in one of the most fundamental of gospel truths? Supposedly, at least according to many Bible teachers, one of the most fundamental truths about the rapture rescue of believers which could ever be taught anywhere by anyone is the idea that the rapture can happen at *any-moment without any intervening events required*. If that is what Paul taught, why would he throw everything out the window and start believing some strange angel who was contradicting the truth of the gospel of rapture imminency? Had the Apostle gone berserk?

Instead of doubting this angel, the Apostle indicated he believed the angel as even a superficial reading of Acts chapter twenty-seven will indicate.

> *23 For there stood by me this night the angel of God, whose I am, and whom I serve,*
>
> *24 Saying, Fear not, Paul; thou must be brought before Caesar: and, lo, God hath given thee all them that sail with thee.*
>
> *25 Wherefore, sirs, be of good cheer: for I believe God, that it shall be even as it was told me.*
>
> *26 Howbeit we must be cast upon a certain island. (Acts 27:23-26)*

Notice particularly verse twenty-five. Paul states, "I believe God, that it shall be even as it was told me."

Hold everything! Paul didn't say he had seen God. He specifically stated he had seen an angel of God. How is it that the Apostle can start equating the words of some strange angel as a message from God? Didn't Paul practice his own preaching? If an angel even appears and contradicts the already preached gospel, reject the angel's words. But does Paul practice that here?

Is it possible that Paul never taught an *any-moment rapture requiring no intervening events?* Is it possible that the popular opinion of imminency rapture supporters is actually incorrect about Paul's gospel message concerning the rapture? If the Jew from the tribe of Benjamin was really an "imminent rapture" teacher and preacher, wouldn't he have known better than to accept a perverted rapture gospel which suggested a rapture was no longer in the works until Paul could eventually get to Rome? If this tentmaker was really an any-moment rapture preacher and teacher who denied that intervening events could occur, why would he gullibly fall for the message of this strange angel who was now proclaiming a prolonged postponement for a rapture during Paul's lifetime? Could it be that Paul's message about the rapture didn't really quite conform to the popular opinion of today's *any-moment, imminent rapture advocates?*

There is probably another reason Paul accepted this angel's message, other than the desperation of the moment. The angel's message was actually quite in harmony with a message delivered by the Messiah Himself to the Apostle on a previous occasion.

> *11 And the night following the Lord stood by him, and said, Be of good cheer, Paul: for as thou hast testified of me in Jerusalem, so must thou bear witness also at Rome. (Acts 23:11)*

In this passage of Scripture it is the Lord Himself who stands beside Paul and spoke words of comfort and assurance to him in very desperate circumstances during his recent imprisonment. There the Messiah Himself denied that an *any-moment, imminent rapture requiring no intervening events* was a hope for Paul's life before he got to

Rome. The Messiah Himself had postponed the rapture for a lengthy period of time in Paul's life. How could the Apostle during this intermediate time before his arrival at Rome be accused of teaching a lie? How could anyone accuse Paul of falsely teaching something he himself did not even believe? Could Paul as a man of integrity teach an *any-moment rapture gospel requiring no intervening events* before his arrival at Rome when he himself knew the very concept was a lie before he got to Rome?

Paul did not believe in the any-moment, imminent rapture gospel as is widely taught today. Paul had divine and supernatural assurance that the rapture was not going to occur before he got to Rome. Paul had supernatural knowledge contradicting the idea that the rapture could happen at any-moment during his lifetime. Both Jesus and an angel of God had provided a double witness that the modern theory of an any-moment, imminent rapture was a false teaching for at least an extended period of his life.

As we have just observed, there are times when the expected rapture predicted by Jesus and Paul was not a possible event within a specified time frame in earth's chronology. The period of time between Messiah's appearance to Paul, in Acts 23:11, and the moment he would stand before Caesar was a period of time in which the rapture was a preempted and impossible event. The angel who had appeared to Paul aboard the ship he sailed in while on the Mediterranean had confirmed the postponed rapture message the Apostle had previously been entrusted with by the Messiah Himself. The message was that the rapture was definitely not on the agenda until Paul stood at Rome.

If the rapture were to have occurred before Paul saw Rome, Paul could not have stood before Caesar. If Paul had been raptured before he had gotten to Rome, he would have been caught up into the clouds to be with the Lord and he could never have gotten to Rome. If Paul had been raptured before he had gotten to Rome, he could never have stood trial before the Roman Emperor. God's plan for Paul to testify before Caesar had made the rapture an impossible event until the predictions by Jesus and the angel of God to Paul could be fulfilled.

The Apostle was confident that he would stand before Caesar, and so he was also confident the rapture would not occur before the predictions by the Messiah and the angel of God could be fulfilled. The words of the Messiah Himself, and the words by the angel of God reminded the Jew from Tarsus that the rapture was not planned for the intervening period of time. If the rapture had taken place during that intervening period in Paul's life, the Messiah would have been turned into a liar and false prophet, and the angel of God would have been turned into a demon.

Evidently, it is possible to know at times when the rapture will not occur. Paul the Apostle had such knowledge. Is it possible for others as well to know when the rapture will not occur?

The Pharisee of Benjamin's tribe knew for a certainty the rapture could not occur during a specified period of time in his own life. This indicates that the doctrine of an any-moment rapture has not been an accurate doctrine throughout the history of the church, if it even can be considered accurate now. The book of Acts indicates that there have been periods of time in the history of the church when the rapture could not occur at any-moment.

God had revealed to Paul through a personal appearance of

the Messiah, and confirmed through the personal appearance of an angel of God, that there was a definite delay for the rapture by informing him he would stand at Rome at some point in the future.

The blessed hope, the snatching away of Messiah's followers from earth, would have interfered with God's plan for the Pharisee from Tarsus to testify before the Roman Emperor. In order for the Apostle to testify before the Roman Emperor, God would have to delay or postpone the rapture until Paul could fulfill the Lord's plans for him to testify before Caesar. A delay for the rapture had been factored into the life of the Apostle. The plans the Lord had for evangelizing the city of Rome would make it necessary for God to delay a rapture.

The very act of informing the Apostle of the plan to send him to Rome to testify before Caesar, was an indication that God was not about to support the idea of an any-moment, imminent rapture during at least this segment of Paul's life. God had indicated to Paul that the rapture was not to be a part of the Apostle's immediate future. A definite delay for the rapture was fully sanctioned by the supernatural appearances of the Messiah and the angel of God. Paul was divinely informed that his mission would be to "occupy." The appearances of Jesus and the angel of God to Paul would discourage any mistaken notions on Paul's part that a rapture might occur at any-moment. Delay for the rapture was part of God's divine plan, and Paul was fully informed that this rapture delay was divinely planned so that the gospel could reach Rome by way of a specially appointed messenger.

Is Divine Delay Normal?
Some may be tempted to suggest that Paul's special relationship with the Lord was the basis for the "privileged"

information he had about a delay in the rapture. We have already examined other examples in the Scriptures which seem to indicate a delay for the blessed hope. We have also examined instances for many of the early disciples and for all of the other Apostles where delay was part of the divine plan. Not only were there specific periods in the lives of all the Apostles when a rapture was impossible, but we have, in previous chapters, examined passages of the Scriptures delivered through Paul and Jesus which indicate observable delay through unfulfilled signs which can be witnessed by all believers.

A rapture delay has been the normative experience for all believers since the early church. In the next chapter we will examine another example of delay.

Chapter Twelve

Stretch Forth Thy Hands

It must have been a severe blow when Peter heard those words. The Messiah had just predicted how Peter would die.

How would you react if you were told by the Lord the manner in which you were going to die? Perhaps few people look forward with real anticipation to the day of their death. The idea is so remote for many of us we are prone to think about it in an abstract manner. A few people who are terrified at the prospect of death have arranged for their bodies to be frozen after their demise, in hopes of being revived when science has hopefully advanced far enough to restore to them life.

Peter was an apparently impetuous man who spoke his thoughts with little hesitation. At the trial of Jesus, he turned out to be a man who was vacillating and cowardly. Peter denied his Lord three times in one night while Jesus was in custody. Earlier Peter had professed his unwavering loyalty. He vowed to Jesus that he would even die with Him. Peter's declaration of allegiance was made to Jesus personally after the Messiah had predicted to the disciples privately His own impending death.

31 Then saith Jesus unto them, All ye shall be offen-

ded because of me this night: for it is written, I will smite the shepherd, and the sheep of the flock shall be scattered abroad.

32 But after I am risen again, I will go before you into Galilee. (Matthew 26:31-32)

In this prophecy concerning His own death, Messiah alluded to the prophet Zechariah who had made a prediction concerning Jesus. From Zechariah's prophecy, Jesus knew He had to die. In the foreknowledge and plan of the Father, Jesus had to die.

Peter's response was one of fierce loyalty. He declared he would never be disillusioned with or offended by what happened to Jesus. Peter rejected the prediction about his own future behavior as foretold in the prophet Zechariah.

33 Peter answered and said unto him, Though all men shall be offended because of thee, yet will I never be offended. (Matthew 26:33)

There is nothing we know so little of as ourselves–nothing we see less of than our own weakness and poverty. The strength of pride is only for a moment. Peter, though vainly confident, was certainly sincere–he had never been put to a sore trial, and did not know his own strength. Had this resolution of his been formed in the strength of God, he would have been enabled to maintain it against earth and hell.
(Adam Clarke, n.d., Matt.26:33)

Peter contrasts himself with his fellow-disciples. Though they all should fall away, he, at any rate, would remain steadfast. He could not endure to be

included in the 'all ye' of Jesus' warning... and as for failing 'this night,' he will never at any time... be offended in Christ.
(Spence & Excell, n.d., Matt. 26:33)

Peter's rejection of the Zechariah prophecy concerning himself prompted a response from Jesus who was evidently given a prophecy through the Holy Spirit concerning Peter's great claim to courageous loyalty. Peter declared he would even die with Jesus, if necessary.

In response to Peter's claim of resolute allegiance and unwavering faith to Jesus, an allegiance which would even go the the point of dying with the Messiah, Jesus predicted that Peter's declaration of courage and loyalty would evaporate into nothingness that very night, within hours. Jesus, by the Holy Spirit, predicted Peter's cowardly denial by this man of good intentions. The denial would not occur once, but three times.

34 Jesus said unto him, Verily I say unto thee, That this night, before the cock crow, thou shalt deny me thrice.

35 Peter said unto him, Though I should die with thee, yet will I not deny thee. Likewise also said all the disciples. (Matthew 26:35)

'He thought he was able,' says Augustine, 'because he felt that he wished.' The other apostles made a similar assertion, and Jesus said no more, leaving time to prove the truth of his sad foreboding.
(Spence & Excell, n.d., Matt. 26:35)

After the resurrection, Peter and the disciples had another

encounter with Jesus on the shores of Galilee. Jesus told the disciples the very night of Peter's betrayal that He would see them after the resurrection, in Galilee.

32 But after I am risen again, I will go before you into Galilee. (Matthew 26:32)

Already Jesus had appeared to them in Jerusalem, but Jesus had said that He would precede them to Galilee the night of the Last Supper. Here on the shores of the Sea of Galilee, which is also called the sea of Tiberias, the disciples have another confrontation with Jesus.

1 After these things Jesus shewed himself again to the disciples at the sea of Tiberias; and on this wise shewed he himself.

2 There were together Simon Peter, and Thomas called Didymus, and Nathanael of Cana in Galilee, and the sons of Zebedee, and two other of his disciples.

3 Simon Peter saith unto them, I go a fishing. They say unto him, We also go with thee. They went forth, and entered into a ship immediately; and that night they caught nothing.

4 But when the morning was now come, Jesus stood on the shore: but the disciples knew not that it was Jesus.

5 Then Jesus saith unto them, Children, have ye any meat? They answered him, No.

6 And he said unto them, Cast the net on the right side of the ship, and ye shall find. They cast therefore,

> *and now they were not able to draw it for the multitude of fishes.*
>
> *7 Therefore that disciple whom Jesus loved saith unto Peter, It is the Lord. Now when Simon Peter heard that it was the Lord, he girt his fisher's coat unto him, (for he was naked,) and did cast himself into the sea.*
>
> *8 And the other disciples came in a little ship; (for they were not far from land, but as it were two hundred cubits,) dragging the net with fishes. (John 21:1-8)*

This time Jesus also had some predictions for Peter. It is here that Messiah predicted how Peter would die. At the Last Supper Peter had professed his allegiance to the Lord, guaranteeing he would even risk his own personal safety to the very death. Jesus had at that time predicted Peter would deny Him three times. After dining with the disciples, Jesus began talking to Peter. It was during this conversation that Jesus predicted that Peter would die. Jesus was indicating that Peter would have the opportunity to fulfill his previous vow of allegiance to the very death.

> *18 Verily, verily, I say unto thee, When thou wast young, thou girdest thyself, and walkedst whither thou wouldest: but when thou shalt be old, thou shalt stretch forth thy hands, and another shall gird thee, and carry thee whither thou wouldest not.*
>
> *19 This spake he, signifying by what death he should glorify God. And when he had spoken this, he saith unto him, Follow me. (John 21:18-19)*

In this prediction, Peter was informed about the circumstances in which he would die. He would be crucified as an elderly man.

> ... it was a custom at Rome to put the necks of those who were to be crucified into a yoke, and to stretch out their hands and fasten them to the end of it; and having thus led them through the city they were carried out to be crucified.
> (Adam Clarke, n.d., John 21:18)

> He says, 'Wither thou wouldest not,' with reference to the natural reluctance of the soul to be separated from the body; an instinct implanted by God to prevent men putting an end to themselves.
> *Chrysostom*
> (Aquinas, 2000, John 21:18)

> 'Follow me.'] Whether our Lord meant by these words that Peter was to walk with him a little way for a private interview, or whether he meant that he was to imitate his example, or be conformed to him in the manner of his death, is very uncertain.
> (Adam Clarke, n.d., John, 21:19)

Jesus had predicted Peter's death, and then He exhorted the fisherman to follow. Could this have been an allusion to Peter's previous cowardice at the three denials? Could it be that Jesus was giving an opportunity to Peter to gain a new self respect? Perhaps the inference was: Peter, you denied me three times, but you can show your courage and faith by deciding to follow me now, since I have just predicted to you how you will die if you do follow me.

The decision Peter would make now at the exhortation to follow Jesus would determine the nature of his real character. Peter had proven himself a coward three times at the trial of Jesus. Peter had evidently become disillusioned with himself. This disillusionment could have potentially destroyed Peter. Jesus, by predicting the fisherman's death, and by then asking Peter to follow when he now knew what kind of death his discipleship would lead to, provided Peter with the opportunity of demonstrating his real courage and faith to himself and to the world. This would furnish the Apostle with knowledge that would lead to continuous humility, and a reliance upon the strength of the Lord, rather than a false reliance on his own strength and power. Peter would perhaps learn by this where true courage was to be found, in reliance upon God.

Why did Peter have to die? Tradition tells us that Peter demanded to be crucified upside down, deeming himself unworthy to die in the same manner as his Lord. We can rationalize and surmise that Peter had to die to prove his loyalty, but the truth is that all of the Apostles, except John, apparently died as martyrs.

Now it may not take a great deal of brain power to realize that if Peter was going to die, he couldn't be a a living participant in the rapture rescue when the Messiah came to rescue His followers from earth. This inference is often overlooked (Gundry, 1973, p. 39).

We never see in Scripture a statement by Peter that he expected to be a part of the rapture. We do see another possible result of this prophecy concerning Peter's martyrdom as an old man. The following Scriptures may reveal how Peter reacted to the prediction of his death as a senior.

> *1   Now about that time Herod the king stretched forth his hands to vex certain of the church.*
>
> *2   And he killed James the brother of John with the sword.*

3 And because he saw it pleased the Jews, he proceeded further to take Peter also. (Then were the days of unleavened bread.)

4 And when he had apprehended him, he put him in prison, and delivered him to four quaternions of soldiers to keep him; intending after Easter to bring him forth to the people.

5 Peter therefore was kept in prison: but prayer was made without ceasing of the church unto God for him.

6 And when Herod would have brought him forth, the same night Peter was sleeping between two soldiers, bound with two chains: and the keepers before the door kept the prison.

7 And, behold, the angel of the Lord came upon him, and a light shined in the prison: and he smote Peter on the side, and raised him up, saying, Arise up quickly. And his chains fell off from his hands.
(Acts 12:1-7)

Notice the behavior of Peter in this affair. James had just been killed, and Herod, realizing he had gained some fans through this action, had Peter arrested. Herod intended to bring Peter out to the multitude after the passover, evidently to allow the crowds to determine Peter's fate. How did Peter react? He went to sleep. In fact, Peter was sleeping so soundly an angel who had come to rescue him struck him to wake him up! When Peter woke up, he found himself being rescued by an angel, but was still so sleepy he evidently thought he was in the process of having some kind of vision while he was being rescued.

> *8 And the angel said unto him, Gird thyself, and bind on thy sandals.*
>
> *And so he did. And he saith unto him, Cast thy garment about thee, and follow me.*
>
> *9 And he went out, and followed him; and wist not that it was true which was done by the angel; but thought he saw a vision. (Acts 12:8-9)*

While Peter was being rescued from the prison by the angel, he was evidently still so sleepy he believed he was having a vision. When did Peter completely wake up? The following Scriptures seem to reveal when this happened.

> *10 When they were past the first and the second ward, they came unto the iron gate that leadeth unto the city; which opened to them of his own accord: and they went out, and passed on through one street; and forthwith the angel departed from him.*
>
> *11 And when Peter was come to himself, he said, Now I know of a surety, that the LORD hath sent his angel, and hath delivered me out of the hand of Herod, and from all the expectation of the people of the Jews.*
>
> *12 And when he had considered the thing, he came to the house of Mary the mother of John, whose surname was Mark; where many were gathered together praying. (Acts (12:10-12)*

How did Peter react to his apparently intended martyrdom? He went to sleep. Why could Peter sleep so soundly in such dangerous circumstances? Through the whole rescue he was still so sleepy he thought he was having a vision. It wasn't until he was finally outside of the prison, and after the angel was gone, that Peter finally fully realized what had just happened.

Was Peter just a little bit mentally slow? Or did the Messiah's prophecy that he would die as an elderly man have something to do with it? Is it possible that Peter was able to sleep so soundly because he knew he would die as an elderly man?

Jesus had predicted that Peter would die in old age. It does not take a great deal of logic to calculate that if Peter was going to die, he would not be one of the living participants at the time of the rapture. He would be resurrected at the rapture, but he himself would eventually die.

Here is a somewhat more complicated inference from the prophecy by Jesus about Peter's death. It may take some consideration to grasp the implications, but there is another very disturbing inference which can be drawn from the prediction of Peter's death given by Jesus. If Peter had to die, the inference we can logically deduce from the Lord's prediction is that the rapture would not occur in Peter's lifetime. What is even more disconcerting is the next conclusion. If Peter had to die, the rapture could not occur while Peter was alive.

The previously mentioned implications are not the only conclusions which can be drawn from the Messiah's prediction concerning Peter's death as an elderly man. If the rapture couldn't occur while Peter was alive, the teaching of an *imminent, any-moment rapture with no intervening events* was not a valid teaching during Peter's lifetime, it would be a fallacy while Peter was alive. No-one could legitimately expect to be in a rapture during Peter's life span, because the Messiah had predicted Peter's death. We can conclude that Peter's life preempted the possibility of a rapture. We can conclude that Peter's predicted death as an elderly man was an "intervening event" between that time and the rapture, so a *sign-less rapture which could happen at any-moment requiring no intervening events* was a total fallacy from a definitional start to finish during Peter's lifetime. Peter's life and the rapture were mutually exclusive entities.

Yes, there would be a rapture. Scriptures predicted the rapture. Jesus personally predicted the rapture. Paul confirmed there would be a rapture, but Peter would not survive to be a living participant when the rapture finally commenced. Jesus had excluded Peter from the event. The rapture would occur, but Peter himself should not entertain any hope as a living survivor to participate in that future event during his lifetime. Peter himself admits this in one of his letters.

> *13 Yea, I think it meet, as long as I am in this tabernacle, to stir you up by putting you in remembrance;*
>
> *14 Knowing that shortly I must put off this my tabernacle, even as our Lord Jesus Christ hath shewed me.*
>
> *15 Moreover I will endeavour that ye may be able after my decease to have these things always in remembrance. (II Peter 1:13-14)*

In these Scriptures Peter admits that Jesus predicted his death. Peter even claims to expect to die in these passages. Peter was not looking for the "upper-taker" (the Messiah at the rapture), he was looking for the "undertaker" (physical death).

A Divine Perspective and the Rapture

There is another implication from this personal prophecy Jesus had made to Peter. Jesus not only predicted Peter's death, Jesus predicted Peter's death as an elderly man. Scratch the rapture from the expectancy of any living believer who was at all significantly older than that fisherman apostle. Anyone significantly older than Peter would have to have some kind of reason to hope for a significantly longer life span than Peter's, in order to entertain any realistic hope for being a living participant at the time of the rapture. Unless someone who was older than Peter had a realistic hope of outliving the Apostle Peter, they would have to realistically scratch the rapture from their itinerary.

The Greek word which Jesus used to indicate that Peter would be an old man when he died can be transliterated as "gerasees." You may recognize part of the word for "geritol" in the Greek word Jesus used.

Peter may have been at an age comparable to that of Jesus when he walked with the Lord. Perhaps he may even have been a few years older than Jesus. Some believe that Peter was the oldest of the Apostles.

There may be some who would like to salvage the *any-moment, imminent rapture requiring no intervening events* from the dilemma we observe developing in regard to Peter's life expectancy. Some may argue that the gospel of John was the latest of the gospels, and since it alone contains the prediction about Peter's death, the knowledge of this prediction was not necessarily widespread. Most Christians were possibly unaware of this particular prophecy by Jesus which was perhaps primarily intended only for Peter's individual use, and not for public consumption at large. With this mode of reasoning, some might argue that under the circumstances, the doctrine of an *any-moment, imminent rapture requiring no intervening events* was still a legitimate teaching due to the possible ignorance of the Christian populace about this prediction concerning Peter's death. Or these people might even argue that if some Christians did know about the prediction by Jesus of Peter's death, the slow methods of communication in that day would deter many people's immediate knowledge concerning Peter's present condition. With these existing circumstances, some would perhaps argue that Christians might need to resort to an imminent rapture belief because they would not know whether Peter was still alive.

This plea to ignorance focuses on the issue from a human perspective. This appeal to the ignorance of Christians concerning the state of Peter's condition, or even the possible widespread ignorance about the prophecy concerning Peter's death, does appear to make a valid case, from the human point of view, for an "imminent rapture" theory. While this "human perspective" appears to provide legitimate grounds for the teaching of an *any-moment, imminent rapture requiring no intervening events,* the argument really takes the form of rationalizing in order to salvage the idea. This rationalizing approach might salvage the imminent rapture theory for some uncritical thinkers, but the tough question which needs to be made concerns the divine perspective. Wouldn't an *any-moment, imminent rapture requiring no intervening events* be a fallacy if it were believed and Peter were still alive? Even if people were ignorant of the prophecy about Peter's death as an old man, wouldn't the teaching of an *any-moment, imminent rapture requiring no intervening events* be a fallacy if Peter were still alive?

To put it another way, let's look at the situation from the heavenly Father's perspective. 1) Didn't the divine Father know Jesus had predicted Peter's death? Of course He knew. The divine Father is omniscient (all knowing). 2) Didn't the divine heavenly Father know if Peter was still alive? Of course the divine heavenly Father knew at all times whether Peter was dead or alive. The divine heavenly Father is omniscient (He knows everything). 3) Why would the heavenly Father wish to propagate what He knew was a fallacy? In other words, if the divine heavenly Father knew Peter had to die before the rapture could occur, but He also knew that Peter was still alive, why would He want people to promote an *any-moment, imminent rapture requiring no intervening events* to hoodwink people into believing it because of their own ignorance about Peter's predicted death, when He Himself knew the teaching was not truthful at the very moment it was supposedly being taught during Peter's lifetime? Not only did the heavenly Father know an *any-moment, imminent rapture requiring no intervening events* would be untruthful while Peter was alive, Peter himself would know the teaching was inaccurate and untruthful while he himself was alive. In addition, any number of people living at the time who knew about the prediction by Jesus of Peter's death would be aware of the fact the imminent rapture teaching was inaccurate and untruthful while Peter was alive.

It seems less than respectable to picture the heavenly divine Father as actively wishing propagation of an untruthful, any-moment rapture theory which He Himself would know was erroneous, and which Peter and any number of knowledgeable believers would also know was inaccurate during Peter's lifetime. What purpose would the God of truth have in propagating an untruth? The whole concept appears to demean the character and reputation of the God of the Bible. The theory that the revealed Creator of the Scriptures would have a desire for active promotion of an inaccurate and untruthful teaching during Peter's lifetime just does not seem to harmonize with the picture of a God who cannot lie. Notice the following Scriptures.

> *1 Paul, a servant of God, and an apostle of Jesus Christ, according to the faith of God's elect, and the acknowledging of the truth which is after godliness;*
>
> *2 In hope of eternal life, which God, that cannot lie, promised before the world began; (Titus 1:1-2)*

The Apostle Paul in this passage specifically states that God cannot lie. Wouldn't God's alleged desire to promote a false teaching about an imminent, any-moment rapture during Peter's lifetime make God a liar?

> *3 Because I will publish the name of the LORD: ascribe ye greatness unto our God.*
>
> *4 He is the Rock, his work is perfect: for all his ways are judgment: a God of truth and without iniquity, just and right is he. (Deuteronomy 32:3-4)*

Moses calls the Lord the "God of truth and without iniquity." Does this description fit the idea of the divine heavenly Father as a God actively desiring promotion of a teaching which He and Peter and many believers would know was false during Peter's lifetime?

Appealing to people's ignorance as a way of legitimizing an any-moment, imminent rapture teaching during Peter's lifetime demeans the God of the Bible. Rationalizing and claiming an inaccurate and false teaching is truth for some because those people are ignorant about Peter's predicted death does not adequately picture the God of truth which we see declared in Holy Writ. This view appears to picture the Lord as a dishonest "shyster" who is preying on people's ignorance as a method for promoting a false and inaccurate teaching as "truth." This picture of a deity who wishes to propagate an *any-moment, imminent rapture requiring no intervening events* during the very period of time in which Peter's own life was a known intervening event preventing the rapture, somehow fails to portray an honest, truthful, loving deity of noble character.

First Century Imminency
Despite the fact that Jesus predicted Peter's death as an event which would intervene between the time of first century believers and the rapture, there did exist a sort of modified, or conditionally imminent rapture theory which did exist in the early church. This theory is specifically documented in the Scriptures. We find this teaching about a "potentially imminent, any-moment rapture" documented in the Bible. That imminent rapture theory took a somewhat unusual twist, because it more or less appeared not to deny that Peter had to die first.

This modified teaching of an *any-moment, imminent rapture requiring no intervening events* apparently did not deny that Peter had to die. We find this theory documented in the gospel of John. It is hidden away in the very same portion of Scripture alongside the very prophecy which declared that Peter would die.

> *18 Verily, verily, I say unto thee, When thou wast young, thou girdest thyself, and walkedst whither thou wouldest: but when thou shalt be old, thou shalt stretch forth thy hands, and another shall gird thee, and carry*

thee whither thou wouldest not.

19 This spake he, signifying by what death he should glorify God. And when he had spoken this, he saith unto him, Follow me.

20 Then Peter, turning about, seeth the disciple whom Jesus loved following; which also leaned on his breast at supper, and said, Lord, which is he that betrayeth thee?

21 Peter seeing him saith to Jesus, Lord, and what shall this man do?

22 Jesus saith unto him, If I will that he tarry till I come, what is that to thee? follow thou me.
(John 21:18-22)

In this passage Peter, after being told in what manner he would die, saw John, the disciple whom Jesus loved, nearby. Impetuous Peter immediately asked about the fate of his fellow disciple. The following quote indicates the intent of the meaning of the passage in which Peter asked Jesus what would happen to the disciple John.

> 'What of this man?' or, How shall it fare with him [John]?... From the fact that John alone of the Twelve survived the destruction of Jerusalem, and so witnessed the commencement of that series of events which belongs 'to the last days,' many good interpreters think that this is a virtual prediction of fact, and not a mere supposition. But this is doubtful, and it seems more natural to consider our Lord as intending to give no positive indication of John's fate at all, but to signify that this was a matter which belonged to the Master

of both, who would disclose or conceal it as He thought proper, and that Peter's part was to mind his own affairs.
(JFB, n.d., John 21:20-21)

Humans are an amazing species. They are filled with optimism that believes in spite of the most difficult odds. What is interesting is that a sort of first century rapture imminency theory sprung up, based on the very details which declared an *any-moment, imminent rapture requiring no intervening events* was not a legitimate idea. Let us consider the details completely in context to get the entire story.

18 Verily, verily, I say unto thee, When thou wast young, thou girdest thyself, and walkedst whither thou wouldest: but when thou shalt be old, thou shalt stretch forth thy hands, and another shall gird thee, and carry thee whither thou wouldest not.

19 This spake he, signifying by what death he should glorify God. And when he had spoken this, he saith unto him, Follow me.

20 Then Peter, turning about, seeth the disciple whom Jesus loved following; which also leaned on his breast at supper, and said, Lord, which is he that betrayeth thee?

21 Peter seeing him saith to Jesus, Lord, and what shall this man do?

22 Jesus saith unto him, If I will that he tarry till I come, what is that to thee? follow thou me.

23 Then went this saying abroad among the brethren,

> *that that disciple should not die: yet Jesus said not unto him, He shall not die; but, If I will that he tarry till I come, what is that to thee?*
>
> *24 This is the disciple which testifieth of these things, and wrote these things: and we know that his testimony is true. (John 21:18-24)*

In the above passage, we recall the original prediction by Jesus indicating Peter's death as an old man by means of crucifixion. Notice the following details which add some interesting historical developments from an apparently broad circulation of Messiah's prediction about Peter's death during the lifetime of the Apostle John.

> *21 Peter seeing him saith to Jesus, Lord, and what shall this man do?*
>
> *22 Jesus saith unto him, If I will that he tarry till I come, what is that to thee? follow thou me.*
>
> *23 Then went this saying abroad among the brethren, that that disciple should not die: yet Jesus said not unto him, He shall not die; but, If I will that he tarry till I come, what is that to thee? (John 21:18-24)*

The rumor began to circulate that, while Peter was definitely going to die, the Apostle John was not going to die! The idea that the Apostle John would not die became a widely circulated belief in the early church. The following discussion is an attempt to recreate the types of ideas which may have surrounded the prediction of Peter's impending death. The following imaginary conversation is an attempt to recreate a possible scenario which may have resulted in the development of a customized, early church imminent rapture theory . The key personalities are fictitiously given the identities of Messianic Followers One, Two, and Three.

Messianic Follower One: "I heard that Jesus predicted Peter would die."

Messianic Follower Two: "Oh really? How did that happen?"

Messianic Follower One: "Peter betrayed Jesus three times while Messiah was on trial, but Peter had told Messiah before the trial he would personally die with Jesus if necessary."

Messianic Follower Two: "I had heard that before."

Messianic Follower One: "After the resurrection, when Peter saw Jesus, Jesus predicted that the fisherman would be crucified."

Messianic Follower Two: "Maybe that was so Peter could prove he really was a real and loyal follower of Jesus after all."

Messianic Follower Three: "When did Jesus predict Peter would die?"

Messianic Follower One: "Jesus told Peter it would happen when Peter was an old man."

Messianic Follower Two: "Really? I happen to be twelve years older than Peter. Maybe that means I will die and not be caught away in the rapture when the Messiah comes to take His followers away from earth. I will probably be buried and not alive for the rapture."

Messianic Follower One: "Well, you never know. After Peter was told he would be crucified, he asked Jesus what would happen to the Apostle John."

Messianic Follower Three: "So Peter is to die by crucifixion? What did Jesus predict would happen to John?"

Messianic Follower One: "Jesus told Peter that if He wanted John to stick around it wasn't any of Peter's business."

Messianic Follower Three: "I recall reading that Jesus before the Mount of Transfiguration said some of the disciples would not die before the Kingdom had come."

Messianic Follower Two: "The Apostle John was the disciple whom Jesus loved. Could that have been an indication that John wouldn't die?"

Messianic Follower Three: "I'd suppose that is exactly what Jesus was predicting. He was telling Peter that it wasn't his affair to be concerned with the Apostle John surviving until the rapture."

Messianic Follower Two: "I am older than Peter by twelve years, but if Peter is going to be crucified, maybe I will outlive him. If I could live beyond Peter's crucifixion, I would still be alive while the Apostle John is alive. I have a chance of being on earth when the Apostle John is caught up at the rapture when Jesus comes for His followers."

Messianic Follower One: "That certainly is possible."

Messianic Follower Two: "If I can just stay in good enough health to live longer than Peter, I may not die after all. I may live to be caught up into the clouds at the Messiah's coming."

Messianic Follower Three: "That is true! The rapture will probably take place after Peter is crucified, and since the Apostle John isn't going to die, if we all outlive Peter, we may all be in the rapture with the Apostle John when Jesus comes to take His followers to be with Him!"

Messianic Follower Two: "Hurray! I have a chance of being in the rapture after all. I may never die."

The above fictitious dialogue attempts to conjecture a reenactment which might explain how the false idea that the Apostle John would never die developed into a widely circulating rumor. The teaching that the Apostle John would never die is refuted by John the Apostle himself in his own account of Messiah's words.

> *20 Then Peter, turning about, seeth the disciple whom Jesus loved following; which also leaned on his breast at supper, and said, Lord, which is he that betrayeth thee?*
>
> *21 Peter seeing him saith to Jesus, Lord, and what shall this man do?*
>
> *22 Jesus saith unto him, If I will that he tarry till I come, what is that to thee? follow thou me.*
>
> *23 Then went this saying abroad among the brethren, that that disciple should not die: yet Jesus said not unto him, He shall not die; but, If I will that he tarry till I come, what is that to thee?*
>
> *24 This is the disciple which testifieth of these things, and wrote these things: and we know that his testimony*

is true. (John 21:20-24)

John emphasized the words which many people had misinterpreted to mean that he would never die. The Apostle may have seen the need to correct the false rumors, because if he did die, and the Messiah hadn't raptured the Apostle, many believers might have ended up rejecting Christianity on the basis of some false rumor.

The rumor that John would never die is fully explainable in light of the teachings of Jesus and Paul the Apostle about the rapture. The Apostle John rejected the idea that he had been predicted by the Lord to Peter as one who would never die. The Apostle John had witnessed the event at which Jesus had spoken to Peter. He would have known better than anyone what Jesus meant. John specifically pointed out that the rumor falsely arose because of the manner in which Jesus answered Peter's question about the way in which the Apostle John's life would end. Jesus refused to tell Peter about the Apostle John's fate.

It is very likely that wishful thinking was involved in the widely circulated rumor which stated that the Apostle John would never die. It is also likely that wishful thinking is heavily involved in the currently circulated teaching of an *any-moment, imminent rapture requiring no intervening events* in our day. Few believers would wish to die. Almost every believer would be happy with the idea of being caught up to be with the Lord in the clouds without ever dying.

Some Biblical interpreters who have seriously investigated the any-moment, imminent rapture theory have been forced to conclude that rapture imminency really isn't a Biblical teaching when all the evidence is finally evaluated. We will examine the Biblical basis for that theory in the final chapter.

Scattered Sheep

If we remember the events which led Jesus to predict that Peter would betray Him three times, there is a possible inference which Jesus did not explicitly explain at the Last Supper with His disciples. As we recall, Jesus referred to a prediction by the Jewish prophet Zechariah. Jesus had alluded to Zechariah's prophecy at the Last Supper, indicating from the prophecy that all of the disciples would flee and leave Jesus alone. Peter had objected to this prediction of his own cowardly behavior, and so Peter told Jesus personally to His face that the Messiah was mistaken. That was how Peter declared his resolute allegiance. Let us reread these Scriptures to refresh ourselves on the details.

31 Then saith Jesus unto them, All ye shall be offended because of me this night: for it is written, I will smite the shepherd, and the sheep of the flock shall be scattered abroad.

32 But after I am risen again, I will go before you into Galilee. (Matthew 26:31-32)

Verse thirty-one is from the prophet Zechariah.

7 Awake, O sword, against my shepherd, and against the man that is my fellow, saith the LORD of hosts: smite the shepherd, and the sheep shall be scattered: and I will turn mine hand upon the little ones. (Zechariah 13:7)

Jesus applied this verse to Himself, as a prediction indicating His own disciples would leave Him that very night. This passage is right next to another verse which clearly predicts a prophecy about the suffering Messiah.

> 6 And one shall say unto him, What are these wounds in thine hands?
>
> Then he shall answer, Those with which I was wounded in the house of my friends.
>
> 7 Awake, O sword, against my shepherd, and against the man that is my fellow, saith the LORD of hosts: smite the shepherd, and the sheep shall be scattered: and I will turn mine hand upon the little ones. (Zechariah 13:6-7)

... "thrusting through" was also a fit retribution on one who tried to "thrust Israel away" from the Lord (Deuteronomy 13. 10); and perfects the type of the Messiah, condemned as a false prophet, and pierced with "wounds betweem His hands." ... The Holy Spirit in Zechariah alludes indirectly to Messiah, the Antitype, wounded by those whom He came to befriend, who ought to have been his 'friends'... the Jews, 'of whom as concerning the flesh He came,' Romans 9.5), but who wounded Him by the agency of the Romans... Expounded by Christ as referring to Himself... Thus it is a resumption of the prophecy of His betrayal... and subsequent punishment of the Jews... For 'smite' (imperative), Matthew 26.31 has 'I will smite.' The act of the sword, it is implied, is GOD'S act... The scattering of Christ's disciples on His apprehension was the partial fulfillment (Matthew 26.31), a pledge of the dispersion of the Jewish nation (once the Lord's sheep, Psalm 100.3) consequent on their crucifixion of Him. The Jews, though 'scattered,' are still the Lord's 'sheep,' awaiting their being 'gathered' by Him...
(JFB, n.d., Zech. 13:6-7)

The above comments, which preceded but anticipated Israel's modern restoration as a nation, indicate that Zechariah's prophecy about the sheep being scattered applied in one sense to the disciples, but in a larger sense to the Jewish nation of Israel which was scattered at Jerusalem's A.D. 70 destruction.

Jesus had elsewhere predicted the destruction of Jerusalem.

41 And when he was come near, he beheld the city, and wept over it,

42 Saying, If thou hadst known, even thou, at least in this thy day, the things which belong unto thy peace! but now they are hid from thine eyes.

43 For the days shall come upon thee, that thine enemies shall cast a trench about thee, and compass thee round, and keep thee in on every side,

44 And shall lay thee even with the ground, and thy children within thee; and they shall not leave in thee one stone upon another; because thou knewest not the time of thy visitation. (Luke 19:41-44)

In the previous passages of Scripture we observe Jesus predicting Jerusalem's destruction. Later in the Olivet discourse we know that Jesus not only predicted events concerning the end of the age and concerning the time of His own coming to set up the Messianic Kingdom, but He predicted events related to the destruction of the Temple and of Jerusalem as well. We know these are part of the Olivet Discourse because the Disciples had asked Jesus specifically these questions in Matthew 24:3.

We are also aware from Luke's version of the Olivet Discourse that Jerusalem's A.D. 70 destruction was part of the series of events Jesus predicted to His Disciples. In addition, we know early believers interpreted the words of Jesus as referring to the destruction of Jerusalem which took place in A.D. 70. This is clearly alluded to in the Biblical account preserved to us concerning Stephen's trial before the Sanhedrin in Acts.

> *13 And set up false witnesses, which said, This man ceaseth not to speak blasphemous words against this holy place, and the law:*
>
> *14 For we have heard him say, that this Jesus of Nazareth shall destroy this place, and shall change the customs which Moses delivered us. (Acts 6:13-14)*

These verses indicate that the destruction of the Temple and of Jerusalem were part of the predictions by Jesus in the Olivet Discourse, for Stephen had preached these ideas.

The Olivet Discourse, if examined for its chronology, appears to present most of the signs the disciples inquired about in chronological order. The clear inference is that, according to a chronological interpretation of the Olivet Discourse, Jerusalem's destruction was to precede the rapture. The unmistakable inference by Jesus was that the destruction of Jerusalem was a sign which would appear before Messiah's appearance in the clouds at the coming for His followers. We know that this is exactly how the chronology of the Olivet Discourse has, to this point, been fulfilled. Jerusalem's destruction did occur in A.D. 70, and the rapture has not yet taken place. This confirms to us that the signs for the Olivet Discourse are probably given in chronological order.

The clear and unmistakable conclusion which can be drawn from the Olivet Discourse, is that one of the signs which would precede the rapture was the destruction of Jerusalem. The rapture was not a possibility while Jerusalem and the Jewish Temple were still standing. The destruction of Israel's capital city, and its national religious monument, the sacrificial Temple, were necessary events which had to precede the rapture. The destruction of Jerusalem and the Temple were intervening events which had to be fulfilled, before the rapture would become possible. These events were predicted by Jesus in the Olivet Discourse.

Paradoxically, the destruction of Israel's Temple in itself implied a long delay for the rapture. Jesus had indicated the Temple would be destroyed, in the Olivet Discourse, but later in the same Olivet Discourse Jesus predicted the abomination of desolation had to occur before the rapture. So Jesus indicated two things about the Temple in the Olivet Discourse: 1) the Temple would be destroyed before the rapture, 2) the abomination of desolation predicted by Daniel had to occur in the Temple before the rapture. For centuries there has been an attempt to reconcile these two different predictions concerning the Jewish Temple. Many have tried to fit everything into the A.D. 70 destruction of the city and its Temple. The obvious and logical conclusion which can be made is that since A.D. 70 did not fulfill the abomination of desolation predicted by the prophet Daniel, and as predicted by Jesus, the Temple would have to be rebuilt after its predicted destruction had taken place.

The destruction of Israel's Temple, and the rebuilding of Israel's Temple, are both implied in the Olivet Discourse. The non-existence of Israel's Temple between these two predicted signs involving the Temple is clearly an intervening event which is implied in the Olivet Discourse. The non-existence of the Temple, which had to occur after the Temple was destroyed, had to occur before the rapture. We are presently experiencing one of the implied signs in the Olivet Discourse, the non-existence of the Jewish Temple. During the current time of the non-existence of the Jewish Temple the rapture cannot take place. The rebuilding of the Temple is required to make the abomination of desolation predicted by Daniel a possible event. The abomination of desolation is a sign which precedes the rapture in a chronological interpretation of the Olivet Discourse, and it is an event which precedes the rapture according to the Apostle Paul.

In the next chapter, an examination of a well known verse of Scripture will be made which suggests that an *any-moment, imminent rapture requiring no intervening events* is a theory resulting from misguided interpretations of the Bible. The Scriptural data suggests there are concrete evidences which can help believers determine the proximity, the nearness or farness, of the rapture event. The well known Scripture in the next chapter clearly contradicts the concept of an any-moment, sign-less rapture.

Chapter Thirteen

The Heavenly Synagogue Meeting

Have you ever had a friend or relative you have tried to get to go to a Bible study? Maybe they have gotten lax in their habit of gathering with believers. Maybe that person was you. A concerned friend or relative may have prodded you, or maybe you have prodded someone else to attend a Bible centered meeting. In these kinds of circumstances where coaxing some infrequent attender occurs, it is very often the case that a famous Bible verse will enter the picture. Someone will start quoting a famous verse as a method of persuading the reluctant gatherer to resume getting together with believers.

Why should a person who doesn't feel like going to a Bible study or worship service go to one? What valid reason is there for attending some meeting with a group of people who believe the Bible?

As a commotion develops over whether attending a Bible centered event is all that worthwhile, often the attempted persuader will pull out that famous passage and begin quoting it to hopefully generate some incentive on the part of a reluctant gatherer. Let us examine it.

> *24 And let us consider one another to provoke unto love and to good works:*

> *25 Not forsaking the assembling of ourselves together, as the manner of some is; but exhorting one another: and so much the more, as ye see the day approaching. (Hebrews 10:24-25)*

This famous passage of Scripture is literally loaded with dynamite. Not only does this passage urge believers to regularly assemble, it has powerful prophetic implications which I have never seen explored by any Bible teacher or believer.

Who wrote the book of Hebrews? The identity of the author is uncertain. Some believe the Apostle Paul wrote it. Others believe it was written by someone else. A few people have even tried to claim it was written by a woman.

The identity of the writer of Hebrews is uncertain. Certainly there are some passages which appear very similar to those in Paul's letters, yet the author of Hebrews, unlike the Apostle Paul, fails to identify himself. In addition, the arrangement of Hebrews is unlike anything in one of Paul's other letters. Paul affectionately had the habit of referring to Timothy as his "son." Timothy was not Paul's biological son. Timothy was regarded by the Apostle as his spiritual son. Paul did occasionally refer to Timothy not only as a spiritual "son," but also referred to Titus as his spiritual "son." The following references indicate the "son" metaphor is used by the Apostle Paul in direct address to the person being referred to as a "son."

> *2 Unto Timothy, my own son in the faith: Grace, mercy, and peace, from God our Father and Jesus Christ our Lord.*
>
> *18 This charge I commit unto thee, son Timothy,*

according to the prophecies which went before on thee, that thou by them mightest war a good warfare; (I Timothy 1:2, 18)

2 To Timothy, my dearly beloved son: Grace, mercy, and peace, from God the Father and Christ Jesus our Lord. (II Timothy 1:2)

4 To Titus, mine own son after the common faith: Grace, mercy, and peace, from God the Father and the Lord Jesus Christ our Saviour. (Titus 1:4)

Paul also referred to Timothy as a "brother," when not addressing him directly.

1 Paul, an apostle of Jesus Christ by the will of God, and Timothy our brother, unto the church of God which is at Corinth, with all the saints which are in all Achaia: (II Corinthians 1:1)

In Hebrews Timothy is mentioned, and since he is not directly addressed, if Paul was the author, it is understandable why Timothy is not addressed as a "son."

23 Know ye that our brother Timothy is set at liberty; with whom, if he come shortly, I will see you. (Hebrews 13:23)

The Western and Eastern churches have historically divided over the authorship of Hebrews, with the Eastern churches accepting the letter as Pauline, and the Western church historically reluctant to accept it as one of Paul's epistles. Hebrews is addressed to Jewish believers, but Paul was known as the Apostle to the Gentiles.

Regardless of the human author's identity, the Greek indicates it was authored by a male because he uses a masculine case for one of the words. Some have thought Hebrews was written by Barnabbas. Martin Luther suggested Apollos as the author. There has been a lack of unity since the first century on identifying the original writer.

The Day
The major issue which is of concern to us here, is the particular reference the writer of Hebrews makes to an approaching day.

> *25 Not forsaking the assembling of ourselves together, as the manner of some is; but exhorting one another: and so much the more, as ye see the day approaching. (Hebrews 10:25)*

Of what day in particular is the writer of Hebrews making reference to? Probably some would suggest the writer is referring to the "the Lord's day." The term "Lord's day," while not specifically stated, is possibly inferred. The term in New Covenant times could conceivably have come to mean the first day of the week, the day on which Jesus was resurrected, Sunday. Earlier it might have meant the Sabbath. It appears rather unlikely that the writer of Hebrews was referring either to the Sabbath day or to Sunday. The meaning of the passage would appear almost to be inconsequential if the author had been referring to the Sabbath or Sunday, the weekly meeting time of believers. Consider the passage as if it were referring to the Sabbath or to Sunday.

> *25 Not forsaking the assembling of ourselves together, as the manner of some is; but exhorting one another: and so much the more, as ye see the (Sabbath) day (or*

*first day of the week, Sunday) approaching.
(Hebrews 10:25)*

Jewish believers were being addressed by the writer of Hebrews. These were Jewish believers who had suffered persecution and loss of property for their faith in the Messiah. It seems rather unlikely that the Sabbath day, or Sunday, would be addressed in such concerned language by the writer of Hebrews. Some have interpreted the passage to suggest that the "exhorting" took place at regular weekly meetings, such as on the Sabbath (Saturday) or on Sunday. In addition, the context of the verse suggests important end-time prophetic events are being referred to by the author.

24 And let us consider one another to provoke unto love and to good works:

25 Not forsaking the assembling of ourselves together, as the manner of some is; but exhorting one another: and so much the more, as ye see the day approaching.

26 For if we sin wilfully after that we have received the knowledge of the truth, there remaineth no more sacrifice for sins,

*27 But a certain fearful looking for of judgment and fiery indignation, which shall devour the adversaries.
(Hebrews 10:24-27)*

The writer appears to be referring to some end-time event of significant importance as an incentive for assembling together as believers more frequently. It appears to many Bible students that the approaching day is a reference to what has been commonly called by Old

Covenant writers, "the day of the Lord."

The "day of the Lord" had connotations of judgment associated with it, and was believed to be possibly a lengthy period of time in which God dealt with human sin. The phrase appears almost twenty times in the Old Covenant Jewish Scriptures.

The "day of the Lord" is mentioned by the Apostle Paul in II Thessalonians 2:1-5. Some Greek manuscripts for that passage in verse two preserve a word which causes the passage to be read as "day of Christ." The King James Version uses a text which preserves that reading.

> 2   *That ye be not soon shaken in mind, or be troubled, neither by spirit, nor by word, nor by letter as from us, as that the day of Christ [day of the Lord] is at hand. (II Thessalonians 2:2)*

More recent translations prefer another Greek rendering which preserves a phrase more in harmony with the Old Covenant Scriptures, "day of the Lord." The manuscripts preferred by the King James Version give the phrase "day of the Lord" a New Covenant twist or meaning. If the phrase "day of the Lord" is referred to as the "day of Christ," there are connotations which are implied which might not be implied by the Old Covenant version of the phrase. Since the Messiah Jesus is the Lord, the phrases "day of Christ" and "day of the Lord" mean essentially the same thing. The essential meaning has not been lost, but there is a New Covenant inference with the phrase, "day of Christ."

Some Biblical interpreters today attempt to distinguish between the "day of the Lord" and the day of the rapture.

Some Biblical interpreters claim the two days do not overlap, and that the two days are completely separate and distinct from each other. That assertion would be particularly difficult to prove in light of the manuscripts on which the King James Translation is based. Since the King James Version in this passage is based on a line of manuscripts which are perhaps more interpretive, and less accurate to the original text, it is probable that the "day of Christ" can be construed to be the "day of the rapture." Close inspection of the text in II Thessalonians 2:1 will also reveal that Paul is referring to "our gathering unto" Jesus, which is the rapture.

> *1 Now we beseech you, brethren, by the coming of our Lord Jesus Christ, and by our gathering together unto him, (II Thessalonians 2:1)*

That "gathering unto" Jesus is almost universally recognized among dispensationalists to be the rapture. Paul in II Thessalonians 2:1 seems to be referring to the gathering to Jesus in the clouds as mentioned by the Messiah to His followers in the Olivet Discourse.

> *31 And he shall send his angels with a great sound of a trumpet, and they shall gather together his elect from the four winds, from one end of heaven to the other. (Matthew 24:31)*

> *27 And then shall he send his angels, and shall gather together his elect from the four winds, from the uttermost part of the earth to the uttermost part of heaven. (Mark 13:27)*

Paul in II Thessalonians 2:1 seems to be alluding to the rapture, because in that passage he uses a Greek root word preserved in the Olivet Discourse which is used by both Matthew and Mark in preserving the teachings of Jesus about the rapture. Rather than define the "day of the Lord" as a period of time which includes the seven years of turmoil mentioned in the book of Revelation, and as a period of time which excludes the rapture, there appears to be a more appropriate definition. II Thessalonians 2:1 appears to define the "day of the Lord" as a period of time which begins with our "gathering unto" Jesus, the rapture.

> *1 Now we beseech you, brethren, by the coming of our Lord Jesus Christ, (appearance of Messiah in the clouds at the rapture) and by our gathering together unto him, (the catching up of living believers to be with the Messiah in the clouds at the rapture)*
>
> *2 That ye be not soon shaken in mind, or be troubled, neither by spirit, nor by word, nor by letter as from us, as that the day of Christ [the day of the Lord] (the day of the rapture rescue) is at hand.*
> *(II Thessalonians 2:1-2)*

The writer of Hebrews is evidently referring to the "day of the Lord." It would appear that the writer of Hebrews, by implying the phrase "day of the Lord," is referring to the rapture which the Apostle Paul referred to by the same phrase in II Thessalonians 2:1-2.

There is a common teaching circulating among a certain segment of dispensational thinkers today which states that the rapture, the snatching away from earth of Messiah's followers, is an event which will occur without any signs. Hebrews 10:25 appears to suggest otherwise.

> *25 Not forsaking the assembling of ourselves together,*

as the manner of some is; but exhorting one another: and so much the more, as ye see the day (of the Lord, the day of judgment beginning with the rapture) approaching. (Hebrews 10:25)

The Signed but Sign-less Rapture
Several years ago I had the opportunity to hear one of America's most prominent pre-tribulational rapture teachers during the previous century. I will not mention his name, for he is now deceased. I have since seen an explanation by another pre-tribulational rapture supporter who related the same explanation I heard the now deceased scholar present at a meeting I was fortunate enough to attend. The more recent living writer referred to the same explanation I had heard the deceased scholar use. The same explanation I had heard was also given by the living writer who stated that his illustration was given by someone whom he did not name. The person I had heard give the same explanation, was, more than likely, the same one the other writer had referred to anonymously.

The explanation I am referring to concerned the paradoxical statements made by some who state the rapture is a "sign-less event," but who will often then contradict themselves and state the signs are all shaping up more clearly indicating the Lord could come very soon. This almost sounds like double speak. The late scholar stated that while he considered the rapture to be an unquestionably sign-less event, that wasn't the complete story. As a major pre-tribulational leader, that deceased scholar made a strong distinction between the rapture, which is the snatching up into the clouds of living believers, and the subsequent event, the later return of Messiah to earth. He believed that seven years minimum separated the two events.

In attempting to reconcile the apparently contradictory statements about the rapture as being an event which did not have any signs, and the statements by many Biblical interpreters that events are more clearly arranging themselves for the coming of the Lord, he used the analogy of two famous holidays: Thanksgiving, and Christmas. He compared the Thanksgiving Holiday to the rapture, which he defined as a sign-less event which would take place at least seven years prior to the future return to earth of Jesus. He further compared Messiah's return to earth to the Christmas Holiday. The late author then stated that as one lives through events during the course of a year, a person may be waiting for the Thanksgiving Holiday to arrive. If the Thanksgiving Holiday has not yet arrived, and a person begins to notice Christmas decorations and advertisements, that person would be in a position to assume that the Thanksgiving Holiday is about to occur very soon, because on the calendar, Thanksgiving chronologically falls before Christmas. The deceased scholar made the analogy that the signs we see today in the world are signs for Messiah's return to earth after the seven years of turmoil in the book of Revelation. These signs, he stated, are not for the rapture. Like Thanksgiving and Christmas, if Christmas decorations appear, they are not signs for Thanksgiving, they are signs for Christmas. In the same way, he maintained there are no signs for the rapture, only signs for the return of Jesus to earth at the end of the seven years of turmoil in the book of Revelation.

The deceased scholar made the statement that since the rapture hasn't happened yet, signs which we see for the end of the age are really not signs for the rapture, but they indicate the rapture is closer because they are signs for the bodily return to earth of Jesus which happens after the seven years of turmoil in the book of Revelation. While the explanation using Thanksgiving and Christmas did explain his position, it would appear that the explanation did not support his theory that there aren't any signs for the rapture.

It is obvious that the deceased advocate of a pre-tribulational rapture did agree signs do exist for the rapture, despite his claim that none existed. The scholar was admitting that signs for Christmas are indirectly signs for the Thanksgiving Holiday, if the Thanksgiving Holiday hasn't arrived yet. In the same way that scholar was actually admitting that indirect signs existed for the rapture, although he claimed they were actually for Messiah's return to earth at the end of the seven years of turmoil. If indirect signs exist for the rapture, then signs do exist for the rapture, even if one continues to repudiate the idea that the rapture has signs. He denied direct signs existed for the rapture, but he admitted, by his analogy, that indirect signs did exist for the rapture.

When people continue to claim the rapture is a sign-less event, and then when they admit that indirect signs exist for that sign-less event, it would appear they are fooling themselves about the whole issue, or that they have some deficiency in expressing themselves, or they are playing some semantical game.

Either signs exist for the rapture, or they don't. If indirect signs exist for the rapture, signs still exist, and the rapture does have signs. Obviously, the now deceased scholar was attempting to maintain an any-moment, sign-less rapture theory, while admitting that world events seemed to indicate Messiah's coming rapture event certainly appeared to be a much nearer and more likely possibility than ever before. It appears that the now deceased scholar was more pragmatic about his explanation of that particular issue concerning signs for the rapture, than he was accurate in claiming the rapture as a sign-less event.

While the writer of Hebrews does not directly specify the day he is referring to, he infers that the day was of such significance, it could be observed as something which was approaching. There is very little doubt that he was referring to the same event Jesus referred to in the Olivet Discourse in Matthew 24:31. That event has been previously demonstrated to be one of remarkably detailed similarity with the I Thessalonians 4:16-17 rapture. The writer of Hebrews, like Jesus, seems to recognize that signs do exist for the rapture, and that the proximity of that event, its nearness or farness, can be gauged by related signs observable in the world.

Unfortunately, the results which I have seen among some believers because of a sign-less rapture teaching is the idea that, "Since the rapture is sign-less, forget about it. Just be ready." Certainly that is the most practical approach one could take to the whole situation, if the rapture were indeed sign-less. It would appear to be almost flawless logic to take the ostrich approach of burying one's head in the sand, concerning the signs of the times, if it were actually the truth that the rapture was sign-less. The only problem is that the teaching of a sign-less rapture is not a conclusion based on flawless logic, and the statement is not a flawless description. The teaching that the rapture is sign-less is also not a conclusion based on a flawless interpretation of the Scriptures. In addition, Jesus cautions us against using the ostrich attitude of burying one's head in the sand concerning signs, for Jesus Himself specifically gave an entire list of signs to watch for in the Olivet Discourse when His disciples asked for signs (Matt. 24:1-3).

The writer of Hebrews 10:25 obviously assumes the approaching day to which he is referring is a day which one can measure the closeness or nearness of. To bury one's head in the sand in an attempt to avoid thinking about the nearness of end of the age events, runs contrary to the teachings of Jesus.

*37 And what I say unto you I say unto all, Watch.
(Mark 13:37)*

Jesus specifically told His followers to "watch," and He also stated by implication that all believers of all times should also "watch."

An Intended Pun?
It would appear that the writer of Hebrews is using a pun in Hebrews 10:25. The pun is built around the Greek word for gathering or assembly, *"episunagogee."* You may be able to recognize the word "synagogue" in it.

The same Greek word is used for gathering and assembling in the following different Scriptures. The words have the same root word for synagogue in them. The Greek words which have the same root word for "synagogue" in them will be indicated in the following passages by brackets, and will be identified by the phrase "synagogue assembling."

*25 Not forsaking the assembling (synagogue assembling) of ourselves together, as the manner of some is; but exhorting one another: and so much the more, as ye see the day (of the Lord, the time of judgment beginning with the rapture) approaching.
(Hebrews 10:25)*

1 Now we beseech you, brethren, by the coming of

> our Lord Jesus Christ, (appearance of Messiah in the clouds at the rapture) and by our gathering (synagogue assembling) together unto him, (the catching up of living believers to be with the Messiah in the clouds at the rapture) (II Thessalonians 2:1)

In Hebrews 10:25, the Greek word used for assembling can be transliterated as *"episunagogee,"* which appears only here and in II Thessalonians 2:1. The verb "episunago" is closely related, and has the same root word used for synagogue, and appears in Matthew 24:31 and Mark 13:27. The Greek words which have the same root word for "synagogue" in them will be again indicated in the following passages by brackets, and will be identified by the phrase "synagogue assembly."

> *31* And he shall send his angels with a great sound of a trumpet, and they shall gather (synagogue assembly) together his elect from the four winds, from one end of heaven to the other. (Matthew 24:31)

> *27* And then shall he send his angels, and shall gather (synagogue assembly) together his elect from the four winds, from the uttermost part of the earth to the uttermost part of heaven. (Mark 13:27)

Almost all dispensationalists agree that the "gathering" or "assembling" Paul referred to in II Thessalonians 2:1 is the rapture. The coming of the Messiah in the clouds, and our gathering together unto Him are two aspects of the rapture event. That event, the rapture, is a "synagogue assembling" event.

> *1* Now we beseech you, brethren, by the coming of our Lord Jesus Christ, (appearance of Messiah in the

clouds at the rapture) and by our gathering (synagogue assembling) together unto him, (the catching up of living believers to be with the Messiah in the clouds at the rapture)

2   That ye be not soon shaken in mind, or be troubled, neither by spirit, nor by word, nor by letter as from us, as that the day of Christ [the day of the Lord] (the time of judgment beginning with the rapture) is at hand. (II Thessalonians 2:1-2)

The Greek word containing the root word for "synagogue" is used by the writer of Hebrews in 10:25 That same root word used for "synagogue" is also used by Paul to refer to the rapture in II Thessalonians 2:1. In the Olivet Discourse the same root word used for "synagogue" can be found preserved in Matthew and it refers to a gathering at the rapture. In Mark's version of the Olivet Discourse the same root word used for "synagogue" can be found preserved and it refers to the rapture.

The apparently intended pun in Hebrews 10:25 takes a Greek word Paul uses for the rapture and applies it to earthly gatherings, or synagogue meetings. Evidently the writer of Hebrews in that verse is making a pun which contrasts a heavenly synagogue gathering (the rapture), as a reason to have more frequent earthly synagogue meetings. The idea which appears to be conveyed in Hebrews is that because of the great approaching day of heavenly assembly (at the rapture), earthly assembly by living believers should occur even more frequently.

These assembly events are mentioned in the Olivet Discourse, and appear to be identifiable as the rapture.

31 And he shall send his angels with a great sound of a trumpet, and they shall gather (synagogue assembly) together his elect from the four winds, from one end of heaven to the other. (Matthew 24:31)

27 And then shall he send his angels, and shall gather (synagogue assembly) together his elect from the four winds, from the uttermost part of the earth to the uttermost part of heaven. (Mark 13:27)

These assembly events in Matthew and Mark appear to be the snatching up of believers into the clouds. The writer of Hebrews seems to be urging his Jewish readers to assemble more on earth, as they see the heavenly rapture assembly day approaching ever more closely.

25 Not forsaking the assembling (synagogue assembling) of ourselves together, as the manner of some is; but exhorting one another: and so much the more, as ye see the day (of the Lord, the rapture, synagogue assembling) approaching. (Hebrews 10:25)

Greek and "Synagogue"

The Greek word used in Hebrews 10:25 for assembling is the word *"episunagogee."* Very likely you can immediately see the word "synagogue" in the transliterated word. The English word "synagogue" refers to a Jewish place of worship, or a building used for Jewish worship.

The Greek word *"sunage"* joins the Greek word *"sun"*, which means "with", and the Greek word *"ago"*, which means "go" or "come." Synagogue thus means to "go or come with" or "go or come together." The prefix "epi" in *"episunagogee"* means "on," or "upon," "by," or "at," and refers to place. "*Episunagogee*" is a word used to refer to the rapture. "*Episunagogee*" thus means to go or come together at a place. "*Episunagogee*" can then be translated as "assembling," or "convening," or "gathering."

Below in II Thessalonians 2:1 "*episunagogee*" refers to the rapture, and is translated as "gathering."

> *1 Now we beseech you, brethren, by the coming of our Lord Jesus Christ, and by our gathering ("episunagogee") together unto him, (II Thessalonians 2:1)*

Below in Hebrews 10:25 "*episunagogee*" is translated as "assembling."

> *25 Not forsaking the assembling ("episunagogee") of ourselves together, as the manner of some is; but exhorting one another: and so much the more, as ye see the day approaching. (Hebrews 10:25)*

The verb form of "*episunagogee*" is used in Matthew 24:31 and is translated as "gather together."

> *31 And he shall send his angels with a great sound of a trumpet, and they shall gather (synagogue assembly) together his elect from the four winds, from one end of heaven to the other. (Matthew 24:31)*

The above verse matches the rapture description of I Thessalonians 2:16-17 almost identically.

In Mark 13:27 an account of the same event in Matthew 24:31 is given. It uses the same verb form of the word "*episunagogee*" used in Matthew 24:31.

> *27 And then shall he send his angels, and shall gather (synagogue assembly) together his elect from the four winds, from the uttermost part of the earth to the uttermost part of heaven. (Mark 13:27)*

Mark 13:27 also matches the Thessalonians 2:16-17 rapture description.

The Observable Day
If the author of Hebrews was indeed using a pun or word play on the Greek word frequently used for the rapture, it is obvious he would repudiate the idea that the heavenly rapture day of assembly is an event which absolutely doesn't have any signs. It seems highly probable that the writer of Hebrews did indeed have the Greek word frequently used for the rapture in mind when he wrote his exhortation to Hebrew followers of the Messiah. He also specifically suggests that the day of heavenly assembly at the rapture is a day which can be visibly seen to approach. Not only did the Apostle Paul use the identical Greek word used by the writer of Hebrews in 10:25, for the rapture, but similar words derived from the root Greek word for "synagogue" were used in the Olivet Discourse by Matthew and Mark to refer to the rapture as Jesus explained it. Those Olivet Discourse descriptions by Jesus of the rapture, as we have observed in a previous chapter, have amazing parallels to the Apostle Paul's description of the rapture. In addition, those rapture descriptions by Jesus in the Olivet Discourse have been associated with the rapture described by Paul the Apostle since the days of the early church Fathers.

It is highly probable that Hebrews 10:25 is a pun or word play which uses and has in mind a particular Greek word for the rapture. That Greek word contains the root word for "synagogue" in it. The writer of Hebrews appears to use the word for "assembling" with a double meaning. Not only does the Greek word connote weekly gatherings for teaching, worship, and prayer, the same Greek word also connotes the assembling of believers to the Messiah in the clouds. That heavenly assembly of believers to the Messiah in the clouds has come to be known in modern times as the "rapture."

The writer of Hebrews 10:25 clearly infers by his word pun on the Greek word used for the rapture that the proximity, the nearness or farness, of the rapture day can actually be observed for closeness.

In the following interpretational paraphrase of Hebrews 10:25 I have taken the liberty to amplify in brackets the text to display its apparently intended word pun.

> 25 *Not forsaking the (earthly) assembling of ourselves together, as the manner of some is; but exhorting one another: and so much the more, as ye see the day (of heavenly assembling) approaching.*
> *(Hebrews 10:25)*

It seems very probable in Hebrews 10:25 that when the writer states, "as ye see the day approaching," he is alluding back to the same Greek word in the earlier part of his sentence, "episunagogee," which can mean both earthly assembling for worship, and the heavenly assembling or gathering at the rapture. Hebrews 10:25 seems to be saying, "Have more earthly synagogue meetings, as you see the day of the heavenly synagogue meeting approaching."

The writer of Hebrews evidently believed the day of the rapture had observable signs, contrary to the popular teaching which claims that the rapture is a "sign-less" event.

This is important, because it indicates the idea of an *any-moment, imminent rapture requiring no intervening events* does not really fit the Biblical description as given in the book of Hebrews. The writer of Hebrews implied that the approach of the day of the rapture was something which could be observed. If one believes that the rapture doesn't have any signs, how can its approach be seen? The truth is, the rapture does have signs, and believers can watch that day approach.

In this chapter, examination of a well known verse of Scripture was made which suggests that an *any-moment, imminent rapture requiring no intervening events* is a theory resulting from misguided interpretations of the Bible.

In the next chapter an examination is made to determine whether believers are exempt from the time of the seven years of turmoil recorded in the book of Revelation, as some Biblical interpreters have claimed.

Chapter Fourteen

The Non-Blessed Hope?

There is a song with a beautiful melody which contains the words, "... like a bridge over troubled waters."

How many people enjoy problems? Hopefully most of us as human beings learn to steer clear of the troubling events in life. We as human beings don't relish difficulties or hard times. Most of us probably prefer times without difficulty, and times without problems. At the announcement of trouble ahead, I frequently have the tendency to turn around and head in the opposite direction.

Trouble is not always bad. Problems and difficulties can be blessings. Some of the greatest human achievements were made because of problems and difficulties. Trouble is not always a curse. Sometimes trouble can be a blessing in disguise. Who would have heard of George Washington if there hadn't been any political problems between England and America? Who would have heard of Abraham Lincoln if there hadn't been a slavery issue? Some great business leaders have assumed control of troubled companies and have made names for themselves by turning those problem ridden corporations around. Trouble can sometimes help to build character. Trouble can lead to ingenious solutions, and trouble can sometimes be the remedy for a bad situation.

In Romans 8:28 the Apostle Paul wrote some lines that have probably inspired believers in the Messiah for every succeeding century since those words were written.

> *28 And we know that all things work together for good to them that love God, to them who are the called according to his purpose. (Romans 8:28)*

To understand this verse aright, let us observe: 1. That the persons in whose behalf all things work for good are they 'who love God,' and, consequently, who live in the spirit of obedience. 2. It is not said that all things shall work for good, but that... they work now in the behalf of him who loveth now... for both verbs are in the present tense. 3. All these things 'work together;' while they are working, God's providence is working, his Spirit is working, and they are working TOGETHER with him. And whatever troubles, or afflictions, or persecutions may arise, God presses them into service; and they make a part of the general working, and are caused to contribute to the general good of the person who now loves God, and who is working by faith and love under the influence and operation of the Holy Ghost. They who say sin works for good to them that love God speak blasphemous nonsense. A man who now loves God is not sinning against God; and the promise belongs only to the present time: and as love is the true incentive to obedience, the man who is entitled to the promise can never, while thus entitled, (loving God,) be found in the commission of sin. (Clarke, n.d., Rom. 8:28)

In the following passage which uses the Greek word for tribulation, we see that a by-product of the trouble was a dependance on God.

8 For we would not, brethren, have you ignorant of our trouble which came to us in Asia, that we were pressed out of measure, above strength, insomuch that we despaired even of life:

9 But we had the sentence of death in ourselves, that we should not trust in ourselves, but in God which raiseth the dead: (II Corinthians 1:8-9)

The following comment indicates that the result of this tribulation by Paul and his fellow workers changed their lives.

What they did in their distress; They trusted in God. And therefore they were brought to that extremity, that they should not trust in themselves, but in God... (Henry, n.d., II Cor. 1:8-9)

Armed with this knowledge that trouble in and of itself is not disaster for someone who follows the Messiah, let us consider some neglected passages of Scripture to determine just how exempt believers in Jesus are from the future time period of the seven years of turmoil found predicted in the book of Revelation. Those seven years of trouble are popularly called the "tribulation," and they are the seven years of Daniel's seventieth week.

The Greek Word for Tribulation
In Revelation 7:14 almost all dispensationalists would agree that the word translated "tribulation" refers to some part of the time frame that is part of the seven years of turmoil in the book of Revelation which are Daniel's seventieth week. The Greek word used there for "tribulation" is derived from "thlipsis," which means "pressure" or "affliction" (Young, n.d., tribulation).

> *14 ... These are they which came out of great tribulation, and have washed their robes, and made them white in the blood of the Lamb. (Revelation 7:14)*

Matthew 24:21 is another passage probably most dispensationalists would probably agree upon as referring to some part of the time frame of Daniel's seventieth week, which is predicted in the book of Revelation. The same Greek word for "tribulation" is used.

> *21 For then shall be great tribulation, such as was not since the beginning of the world to this time, no, nor ever shall be. (Matthew 24:21)*

Matthew 24:29 also contains a reference most dispensationalists would likely agree refers to some part of Daniel's seventieth week, the future period of seven years depicted in the book of Revelation popularly called the "tribulation." The same Greek word for "tribulation" appears here.

> *29 Immediately after the tribulation of those days shall the sun be darkened, and the moon shall not give her light, and the stars shall fall from heaven, and the powers of the heavens shall be shaken:*
>
> *30 And then shall appear the sign of the Son of man in heaven: and then shall all the tribes of the earth mourn, and they shall see the Son of man coming in the clouds of heaven with power and great glory. (Matthew 24:29)*

Mark 13:24 is one other passage probably the majority of dispensationalists would agree upon as referring to a period of time within Daniel's seventieth week which are the seven years of turmoil predicted in the book of Revelation.

The same Greek word for "tribulation" is used again.

> *24 But in those days, after that tribulation, the sun shall be darkened, and the moon shall not give her light, (Mark 13:24)*

All of the preceding passages use Greek words for tribulation which are derived from "thlipsis." The same root Greek word is used not only in the four passages just cited, which refer to the seven years of Daniel's seventieth week which are predicted in the book of Revelation, but the same root Greek word is used throughout the New Covenant Scriptures as a word referring to events of a troublesome nature during periods of time outside the book of Revelation.

Non-Tribulation Tribulation
Let us examine some other uses in the New Covenant Scriptures for the same Greek word just examined which do not explicitly apply to the seven years of Daniel's seventieth week predicted in the book of Revelation which is popularly called the "tribulation."

In John 16:33 Jesus gave His disciples a promise.

> *33 These things I have spoken unto you, that in me ye might have peace.*

> *In the world ye shall have tribulation: but be of good cheer; I have overcome the world. (John 16:33)*

This passage uses the same Greek word for "tribulation" used in other passages referring specifically to the seven years of Daniel's seventieth week predicted in the book of Revelation. How many believers have memorized this promise? The Greek word for "tribulation" is the identical word and form used in Matthew 24:29 and Mark 13:24. The latter two verses use the same Greek form of the word used in John 16:33. These gospel uses of the word for "tribulation" in Matthew 24:29 and Mark 13:24 are believed to refer to part of the seven years of Daniel's seventieth week predicted in the book of Revelation. The Greek word in John 16:33 for "tribulation" is from the same root Greek word used for all New Covenant Scripture appearances referring to the time frame of the seven years of Daniel's seventieth week predicted in the book of Revelation. How many believers are aware that Jesus promised His followers turmoil and trouble and tribulation using the same Greek word for "tribulation" which is used elsewhere by Jesus to refer to the seven years of Daniel's seventieth week predicted in the book of Revelation?

Interestingly, all of the Apostles died as martyrs except for the Apostle John. Some were flayed alive. Peter was crucified upside down. Could the future seven years of Daniel's seventieth week produce any more horrendous events than those some of Messiah's immediate Apostles suffered? Since the Scriptures do refer to that future period as the most horrible earth will ever experience, it seems probable that those future years are worst not necessarily because of the quality of trouble, but because of the quantity of trouble. Those future years may include both: *the greatest quantity of quality trouble the earth has ever witnessed in the history of the world.*

In Acts 14:22 we find another instance of the same Greek word for tribulation used to refer to periods within the seven years of Daniel's seventieth week. By considering the context, Acts 14:22 is observed to be a statement made by a group of believers which evidently included the Apostle Paul who had just been stoned, supposedly to death, but who revived, or was resurrected. Subsequently, in passing through several cities these leaders were:

> *22 Confirming the souls of the disciples, and exhorting them to continue in the faith, and that we must through much tribulation enter into the kingdom of God. (Acts 14:22)*

"... we must through much tribulation enter into the kingdom of God." The same root Greek word appears here which is used to refer to the seven years of turmoil.

> Again, as the spirit of the world would be ever opposed to the spirit of Christ, so they must make up their minds to expect persecution and tribulation in various forms, and therefore had need of confirmed souls and strong faith, that, when trials came, they might meet them with becoming fortitude, and stand unmoved in the cloudy and dark day. And as the mind must faint under trouble that sees no prospect of its termination, and no conviction of its use, it was necessary that they should keep in view the kingdom of God, of which they were subjects, and to which, through their adoption into the heavenly family, they had a Divine right. Hence, from the apostle's teaching, they not only learned that they should meet with tribulation, much tribulation, but, for their encouragement, they were also informed that these were the very means which God would use to bring them into his own kingdom; so that, if they had tribulation in the way, they had a heaven of eternal glory as the end to which they were continually to direct their views.
> (Clarke, n.d., Acts 14:22)

Another commentator has made the following remarks about this passage in Acts 14:22.

> They also represented to the believers... that the way to the Kingdom of God, would necessarily... conduct them through many trials. Such instructions and representations, which tended to strengthen their souls, were the more appropriate and necessary, as persecution and affliction might have otherwise perplexed their minds, and induced them to renounce their faith." (Lange, 1866, Acts 14:22)

The Apostle Paul also made the following statement.

> *12 Yea, and all that will live godly in Christ Jesus shall suffer persecution. (II Timothy 3:12)*

Trouble, trials and persecution appear to be part of the package one receives when one follows the Messiah.

A Cursed Hope?
In the desire to persuade others that the pre-tribulation rapture position is the logical position, some promoters have engaged in a method of reasoning which, at first glance, appears to be completely plausible. In applying their logic they use a Scripture from the Apostle Paul's writings.

> *13 Looking for that blessed hope, and the glorious appearing of the great God and our Saviour Jesus Christ; (Titus 2:13)*

Here the appearance of Jesus is described as a "Blessed Hope." Some pre-tribulational rapture promoters focus on this verse and state that the rapture is not a blessed hope if believers are to first expect tribulation, even some of the tribulation which is described in the book of Revelation as the horrible seven years of trouble. These people would claim that if the antichrist must appear first, the blessed hope is not a blessed hope. Some of these pre-tribulational rapture supporters develop this argument and state that the rapture can only be considered a blessed hope if believers are snatched away and rescued completely from the seven years of trouble which are depicted in the book of Revelation. The argument is pushed to assert that if believers enter any part of the seven years popularly called the "tribulation," the rapture cannot be considered a blessed hope. The rapture, supposedly, can only be considered a blessed hope if one can escape the entire tribulation period of all seven years. At first glance, this argument appears to be completely valid.

The rapture, it must be remembered, is the catching up of living believers to be with the Lord. According to I Thessalonians 4:16-17, that snatching away of living believers is immediately preceded by the resurrection of deceased believers.

> *16 For the Lord himself shall descend from heaven with a shout, with the voice of the archangel, and with the trump of God: and the dead in Christ shall rise first:*
>
> *17 Then we which are alive and remain shall be caught up together with them in the clouds, to meet the Lord in the air: and so shall we ever be with the Lord. (I Thessalonians 4:16-17)*

Deceased believers, according to these Scriptures, are part of the event which includes the snatching away of living believers. Dead believers who were faithful to the Messiah will, according to these verses, be resurrected and raised to life prior to any living believer on earth being caught up to be with the Lord. At the same time as the rapture, there is a resurrection. The resurrection takes place first. After dead believers are raised to life, then living believers on earth are caught up to be with the resurrected believers who have already started on their trip to be with the Messiah who is at that moment in the clouds.

Those who claim the rapture can only be considered to be a blessed hope if a snatching away before the seven years of turmoil takes place must have forgotten believers who are dead. What about Peter? Could Peter consider the blessed hope to be a blessed hope when Jesus had already predicted his death by crucifixion? What about all the other Apostles? All the other immediate Apostles of Jesus were martyred except for John. Could they have considered the blessed hope to be a blessed hope while they were going through their martyrdoms? What about the millions of believers who have suffered torture, persecution, and death for their faith. Could they have considered the blessed hope to be a blessed hope?

If Messiah's coming for dead and living believers can only be considered a blessed hope by living believers who expect a pre-tribulational rapture, then neither is that blessed hope a blessed hope for all those who have died. Isolating the blessed hope's blessedness to only those believers who are alive at Messiah's rapture rescue isolates the blessedness to a minority segment of believers throughout the entire history of the church, and deprives the majority of dead believers who are part of that blessed hope from any part of the blessedness of that event.

One can be entirely skeptical of the claim by some pre-tribulational rapture promoters that the blessed hope's blessedness consists in being a minority participant as a living believer, and only if it occurs before the seven years in the book of Revelation popularly called the "tribulation." Notice what Paul states immediately after his description of the rapture in I Thessalonians.

> *16 For the Lord himself shall descend from heaven with a shout, with the voice of the archangel, and with the trump of God: and the dead in Christ shall rise first:*
>
> *17 Then we which are alive and remain shall be caught up together with them in the clouds, to meet the Lord in the air: and so shall we ever be with the Lord.*
>
> *18 Wherefore comfort one another with these words. (I Thessalonians 4:16-18)*

"Wherefore comfort one another with these words." Evidently, Paul was indicating there was a blessedness to the blessed hope which had nothing to do with any pre-tribulational considerations.

The Nature of the Blessedness
The blessed hope's blessedness must be of a nature which is blessed to all participating believers in that event. Both dead and living believers have a right to view the blessed hope as blessed, even if one is staring into the jaws of death, as was Peter at his crucifixion. In order to define the blessed hope's blessednesse, we must recall that many believers throughout history suffered martyrdom, including the Apostle who described the Lord's coming as a "blessed hope." How ironic to believe that the Apostle who wrote

these words himself would be considered ineligible for the blessedness he wrote about.

The blessed hope's blessedness, if it is indeed to be a blessing to all participants, must consist in some value which does not derive its blessedness for a minority of living believers throughout the history of the church. To define the blessed hope's blessedness as escape from the seven years of trouble points to a blessedness which is of only momentary temporal value to a minority segment of believers in the scope of all of eternity.

The real blessedness in the blessed hope consists in a more eternal value which has spiritual worth to all participating members in that event, however they may have passed from this life, and whatever tortures they may have faced in leaving their bodies. The core value in the blessed hope's blessedness lies in the realization that all true followers of the Messiah will finally be conformed to the image of Jesus.

> *2 Beloved, now are we the sons of God, and it doth not yet appear what we shall be: but we know that, when he shall appear, we shall be like him; for we shall see him as he is. (I John 3:2)*

The blessed hope is blessed, because it is the final act in the divine plan of redemption for those who are participants in that event. The blessed hope is blessed to all participating believers in that event, resurrected and living, because it is the moment when mortality puts on immortality.

> *53 For this corruptible must put on incorruption, and this mortal must put on immortality.*

54 So when this corruptible shall have put on incorruption, and this mortal shall have put on immortality, then shall be brought to pass the saying that is written, Death is swallowed up in victory.

55 O death, where is thy sting? O grave, where is thy victory? (I Corinthians 15:53-55)

The blessed hope is blessed for all its participating members, because it is that moment when death, for them, has been swallowed up in victory.

The blessed hope is not blessed because it supposedly brings an escape from seven troublesome years which are only momentary in the panorama of eternity. The blessed hope is blessed because it completes the eternal salvation for those believers who are part of that event. The blessed hope's blessedness doesn't consist primarily in some temporal blessing, but in the eternal completion of the divine plan of deliverance from the effects of sin. That is why the blessed hope is blessed.

When an individual makes a spiritual decision to follow Jesus by faith, the Scriptures reveal a spiritual birth takes place (John 3:3-8). The individual receives a new nature as a result of this spiritual birth (II Cor. 5:17). The new nature is spiritual. The Holy Spirit enters the person's life and that life is changed. The Holy Spirit is a seal, or down payment, for the completion of the work which has begun in that person's life.

13 In whom ye also trusted, after that ye heard the word of truth, the gospel of your salvation: in whom also after that ye believed, ye were sealed with that holy Spirit of promise,

14 Which is the earnest of our inheritance until the redemption of the purchased possession, unto the praise of his glory. (Ephesians 1:13-14)

When the decision to follow Jesus has been made, the body of that believer is not changed. The sinful nature still exists. The redemption of the body is the final act of salvation. The blessed hope's blessedness consists in the fact that the body is finally redeemed, it is finally saved from its evil condition and transformed at the resurrection and rapture which is the blessed hope for followers of the Messiah.

21 Because the creature itself also shall be delivered from the bondage of corruption into the glorious liberty of the children of God.

22 For we know that the whole creation groaneth and travaileth in pain together until now.

23 And not only they, but ourselves also, which have the firstfruits of the Spirit, even we ourselves groan within ourselves, waiting for the adoption, to wit, the redemption of our body. (Romans 8:21-23)

The apostle shows that the whole creation is in a suffering state, into which it has been brought by the disobedience of one man, Adam; therefore, it was made subject to vanity–pain, sickness, and death; not willingly, for mankind had no part in that transgression which 'brought death into the world and all our wo[e];' but God subjected the whole, purposing to afford them a deliverance and infusing into every heart a hope that a more auspicious era should take place; and it is through the influence of this hope, which every man possesses, that the present ills are so patiently borne,

because all are expecting better days. The great deliverer is the Messiah, and the Gospel days the auspicious era which God intended to bring forward. They who believe in Christ with a heart unto righteousness are freed from the bondage of their sinful corruption, and brought into the glorious liberty of the sons of God; and they look forward with joyous expectation, waiting for the... resurrection, when their bodies also shall be redeemed from corruption, and the whole man; body and soul, be adopted into the family of heaven ABOVE, as their souls had been previously adopted into the family of faith BELOW.
(Clarke, n.d., Rom. 8:23)

The blessed hope is blessed, because the last stage of salvation occurs for participants in that event, the redemption of the body.

With this knowledge about the nature of the blessedness which makes the rapture/resurrection a "blessed hope," it is obvious that even if one must pass through some of the seven years of turmoil in the book of Revelation, even a rapture during that period of time can be referred to as a "blessed hope."

Watching for Signs, not Messiah?
Some claim that Messiah's followers must be raptured before the book of Revelation's seven years of turmoil, because if believers are going to spend their time watching for some signs like the antichrist or a rebuilt Temple, they will be watching for signs and not for the coming of Jesus. That is the way some present the argument against a rapture during the seven years of turmoil in the book of Revelation.

Is this a valid argument against a rapture during the seven years described in the book of Revelation? If someone wishes to test the validity of this argument, all they have to do is to gain some practical experience at waiting for someone. Anyone with any experience at waiting for another party or individual will discover that this reasoning is totally fallacious.

As an example, someone I had recently met phoned me and arranged to meet me in approximately fifteen minutes. They were, at the moment they called, a mile or two away. They indicated they would arrive within fifteen minutes, which indicated they would be coming by way of some vehicle. I had no knowledge as to what kind of vehicle they drove. I knew what the individual's appearance was like, but it would have been useless to try to watch for the face of that person in some vehicle, because shadows and sunlight have developed sophisticated techniques for concealing drivers' faces behind windshields.

As the time ticked by, fifteen minutes became twenty minutes. I obviously knew the approximate time the person was expected to arrive, but dozens of vehicles would regularly drive by the location I was at, so I had to look for special signs. If a car or truck slowed down, and the shadowy form of a driver began rotating their head, I would immediately have a strong right to conclude this was my expected visitor.

When the acquaintance did first show up in his vehicle before his U-turn, I could not see his face, but the strange motions of his automobile strongly indicated the driver intended to do some kind of stop in the vicinity. The distance was great enough so that I only caught a glimpse of the driver's face. I never saw his features clearly until he made a U-turn and finally emerged from his vehicle. Before my new acquaintance exited his car, I was already on the sidewalk walking toward his vehicle to greet him, never having known before what kind of vehicle he drove.

It must be noted I was focusing on the behavior of his car. It was the only vehicle I had seen which slowed down and appeared to get ready to stop. He actually slowed down while he was on the other side of the street. That was the first evidence I had concerning the arrival of my expected visitor. He then drove off, and evidently made a U-turn, and was out of sight for a few moments until he came back and his vehicle reappeared on my side of the street. Now I couldn't see his face at all. Without knowing absolutely who it was, I had come to the conclusion only by the observable signs I had seen that this was the individual driving the vehicle I was expecting. Did this mean that I was not looking for my expected visitor because I was focused on the behavior of a car? Absolutely not. Sometimes signs are the only practical means by which a person can watch for an expected visitor in everyday life.

To state that I should have been watching for the person, and not for signs of some vehicle they may have been driving, would have amounted to little more than nonsense. The same is true for the return of the Messiah. To claim that signs would prevent people from watching for Messiah would be like stating because I was watching for a vehicle, I was not really expecting a visitor! Signs in no way impair one's ability to watch for the Messiah. Everyday life verifies regularly that signs are sometimes the only appropriate means by which one is able to watch for an expected visitor. Signs during the seven years of the book of Revelation are an appropriate means to prepare for the appearance of the Messiah which will take place at the rapture.

Not Appointed to Wrath

A Biblical reason often used to support the view that followers of the Messiah are exempt from any tribulation during the seven years of turmoil in the book of Revelation is a passage in I Thessalonians.

> *9 For they themselves shew of us what manner of entering in we had unto you, and how ye turned to God from idols to serve the living and true God;*
>
> *10 And to wait for his Son from heaven, whom he raised from the dead, even Jesus, which delivered us from the wrath to come. (I Thessalonians 1:9-10)*

The argument made is that believers are not appointed to "wrath." "Wrath" is then defined to mean the seven years of turmoil in the book of Revelation. The meaning is then interpreted to be, "Believers are not appointed to spend any time in the seven years of Daniel's seventieth week which are predicted in the book of Revelation."

This argument, which is often advanced to support a pre-tribulational rapture, is rather problematic. Assumptions are made which have a striking similarity to a row of dominoes, where the dominoes are stood on end. If you tip the first domino over, the whole row goes tumbling down.

The whole problem lies in the word "wrath." What is meant by the term? Pre-tribulationalists indicate that the term is an appropriate description for the seven years of the book of Revelation, so this is then drawn out as support for a rapture before those seven dreadful years predicted in the book of Revelation. The strength of this interpretation for a pre-tribulational rapture is that wrath does apparently appear to be a characteristic during those seven years. The great strength in this pre-tribulational rapture argument is also its biggest weakness. Of the almost three dozen appearances of this word for wrath in the New Covenant Scriptures, only six of them appear in the book of Revelation. The majority of the instances for this word outside the book of Revelation, if any, do not refer to the seven dreadful years of Daniel's seventieth week. The majority of the appearances of this word for wrath outside the book of Revelation appear to refer to the time of final judgment.

Consider the same word for "wrath" found in Matthew 3:7.

> *7 But when he saw many of the Pharisees and Sadducees come to his baptism, he said unto them, O generation of vipers, who hath warned you to flee from the wrath to come? (Matthew 3:7)*

Was this referring to the seven years of trouble in the book of Revelation? Obviously not.

There is another similar reference in Luke.

> *7 Then said he to the multitude that came forth to be baptized of him, O generation of vipers, who hath warned you to flee from the wrath to come? (Luke (3:7)*

Consider this following passage from John.

> *36 He that believeth on the Son hath everlasting life: and he that believeth not the Son shall not see life; but the wrath of God abideth on him. (John 3:36)*

Is Jesus referring to the seven years in the book of Revelation? No.

Consider the word for wrath used in Romans.

> *22 What if God, willing to shew his wrath, and to make his power known, endured with much longsuffering the vessels of wrath fitted to destruction:*
> *(Romans 9:22)*

In view of these other New Covenant appearances of the word for wrath or anger, it is highly improbable that I Thessalonians 1:10 or 5:9 have taken on a technical meaning which refers to the seven years of trouble at the end of the Gentile age. Within the same letter by Paul to the Thessalonians, the word for anger or wrath is used again, this time in a way which prevents us from applying the conjectured technical meaning of the seven troublesome years at the end of the Gentile age. Consider the context of I Thessalonians 2:16 where the same word appears, and judge for yourself.

> *14 For ye, brethren, became followers of the churches of God which in Judaea are in Christ Jesus: for ye also have suffered like things of your own countrymen, even as they have of the Jews:*
>
> *15 Who both killed the Lord Jesus, and their own prophets, and have persecuted us; and they please not God, and are contrary to all men:*
>
> *16 Forbidding us to speak to the Gentiles that they might be saved, to fill up their sins alway: for the wrath is come upon them to the uttermost.*
> *(I Thessalonians 2:14-16)*

Obviously, Paul is not suggesting the Jewish nation was currently experiencing the seven years of turmoil at the end of the Gentile era.

The bulk of the occurrences of the Greek word for wrath or anger suggest the probable connotation for the word is that of final judgment in most instances. Attempting to impose an interpretation on the word for anger or wrath, in I Thessalonians 1:10, requiring that the word be interpreted as a reference to the seven troublesome years at the end of the Gentile era appears to be a highly subjective and private interpretation. The ambiguity involved certainly does not allow one to build a doctrinal case for a pre-tribulational rapture on such a subjective scenario. One must assume a pre-tribulational rapture in order to see evidence of a pre-tribulational rapture from the passage. The attempt to find a pre-tribulational rapture in the verse appears to be a desperate attempt to find Biblical support for a rapture rescue before the final seven years of the Gentile era.

Conclusion
This chapter has made an attempt to examine commonly used Biblical passages and resulting arguments which have been advanced to promote the supposed logical requirement for a rapture before the seven years of turmoil at the end of the Gentile era. Some of these arguments have been demonstrated to be fallacious. One argument used by pre-tribulational rapture supporters is highly subjective, and appears to be a desperate attempt to find support for a pre-tribulational rapture in the Bible.
We have also examined a number of Biblical passages which clearly promise trouble for believers. That promised trouble for believers was based on the same Greek word used to refer to the seven troublesome years in the book of Revelation. The Scriptures, and experience, also indicate that times of trouble can sometimes be beneficial. Times of trouble may make us rely on the Lord. As Israel was pre-

served during the ten plagues which fell upon Egypt, the Lord is able to preserve His people through plagues during the seven years at the end of the Gentile era, also. There are passages which do suggest that martyrdom will still take place during those seven years, especially of believers, and that it will be widespread.

The Scriptures seem to suggest a rapture during the last seven years of the Gentile era is entirely in harmony with the Bible. Consider the following text.

> 3 And not only so, but we glory in tribulations also: knowing that tribulation worketh patience;
>
> 4 And patience, experience; and experience, hope:
>
> 5 And hope maketh not ashamed; because the love of God is shed abroad in our hearts by the Holy Ghost which is given unto us. (Romans 5:3-5)

In light of this passage of Scripture, it is evident that trouble can lead to very beneficial developments. Paul declared he gloried in tribulation because of its beneficial consequences.

The following quote indicates that tribulation is not the worst possible event which can happen to one of the Lord's followers.

> 35 Who shall separate us from the love of Christ? shall tribulation, or distress, or persecution, or famine, or nakedness, or peril, or sword?
>
> 36 As it is written, For thy sake we are killed all the day long; we are accounted as sheep for the slaughter.

37 Nay, in all these things we are more than conquerors through him that loved us.
(Romans 8:35-37)

The above passage uses the same Greek word for tribulation also used of the seven years of trouble at the end of the Gentile era. These verses almost sound as if they were written for believers during that future seven years of turmoil. These verses indicate that God's love remains steadfast in the midst of great distress.

This chapter has attempted to demonstrate that the Scriptures do not provide exemption from the complete seven years of trouble at the end of the Gentile era. The pre-tribulational rapture, which is supposedly required by the Scriptures, appears to evaporate when subjected to detailed examination.

In the final chapter an examination will be made of some Bible passages which are perhaps some of the least understood and possibly most misapplied concerning end of the age events in the entire New Covenant Scriptures.

Chapter Fifteen

Sign Language

The child was missing. It was unnerving but true. This time the little boy was not found anywhere. Horror set in as the search party became resigned to the fact that the child was probably now no longer in the land of the living.

He was a little boy fond of fun and games, just like most little boys. He was full of fun, vigor, and life, and had a kind of hysterically funny humor which found pleasure and fulfillment in a morbid joke. He seemed to thrive on something which terrified and frightened others.

Was the child malicious? No. Maybe it was his way of getting the attention he needed and so obsessively desired. People were certainly inconvenienced by his antics and grisly sense of humor, but he never ever intended to do anyone any harm. This time it seems as though the innocent child's humor had pushed the envelope too far.

He pulled his attention getting stunts perhaps to prove to himself that he really was loved and appreciated, despite the nagging doubts which hounded him. He was, perhaps, a classic example of the insecure individual. It probably never occurred to him, or to others, that one day his insecure fetish for attention would finally be his own undoing.

The Planetary Exodus

He, probably least of anyone, thought that there would come a time when he would fall victim to his own prank.

The story of the little boy who cried once too often, "Wolf," is a morality story which has a gruesome and chilling ending, but it genuinely teaches a valuable truth. That truth is that if someone generates a false alarm too many times, people become desensitized to the danger; and the result is that people begin to discredit the message when it is sounded, because it most likely isn't really a valid alarm.

This same kind of dangerous desensitization is exactly what is occurring today among believers who are being taught that the sudden snatching away of Messiah's followers into the clouds can happen at any-moment. The teaching of a *sign-less, any-moment rapture which will happen without any intervening events being required* is a simple case of overkill.

As has been noted in previous chapters, the any-moment pre-tribulational rapture idea may have caught on in modern times possibly as a remedy to the embarrassing Millerite movement which had falsely predicted the exact day for the rapture two years in a row. Anecdotes have been preserved which describe people climbing atop roofs of buildings in hopes of being caught up sooner than those nearer the ground during the unfortunate hysteria generated by William Miller's misguided conclusions concerning the time of the rapture.

Tens of thousands of Bible believing people were subsequently humiliated and embarrassed by the whole affair, whether they had participated or not. Those who had been skeptical of William Miller's prognostications were vindicated, but if they shared the same basic Bible beliefs he had, they would have been party to the humiliation which followed anyway. Many of these more prudent Bible believers were not going to reject the Scriptural predictions of a rapture just because a Bible preacher had made a spectacle of himself and his followers due to his erroneous conclusions.

When Darby began spreading his pre-tribulational rapture theory, the new teaching may have provided a remedy to the false beliefs William Miller had fostered concerning the issue. The "imminency" doctrine was developed by a committee which had attempted to define what probably amounted to an explanation of the Biblical idea of "watchfulness." The committee had taken the "watchfulness" idea, which the Scriptures clearly teach, and had transformed the concept into a very effective and powerful formula which would effectively stop the kind of rapture date-setting William Miller had been guilty of. This new teaching was perhaps designed to be an effective antidote against any potentially similar William Miller debacles in the future. The new formula would effectively prevent individuals from engaging in date-setting shenanigans. Several of the committee members may have had concerns about the widespread humiliation which had accompanied the Millerite controversy.

The new anti-date-setting formula is today known as the teaching of "imminency." It has proven to be extremely effective in discouraging rapture date-setting tendencies by those organizations and religious groups which have adopted it. The "imminency" formula effectively prevents the kind of hysteria which had been generated by Miller's movement.

Setting dates for the rapture has had occasional comebacks

in the previous century (1988, 1989, 1994), and some Biblical interpreters have made calculations for sometime in the first decade of our present century, with possible adjustments being considered on those current rapture calculations because of potential calendar errors.

While an any-moment, sign-less rapture teaching has, for the most part, solved a date setting proclivity, the truth is that a critical analysis of the any-moment rapture theory specifically suggests that while it may have solved a major problem, it itself is an erroneous overcorrection to an incorrect practice.

As has been demonstrated in previous chapters, there existed specific events within the early church which specifically prevented an any-moment rapture. As also previously demonstrated, the "imminency" rapture theory specifically contradicts the idea of contemporary manifestations of the Biblical Charismatic/Pentecostal I Corinthians 12 gift of prophecy as manifested by Agabus and other early believers in a manner which predicts future intervening events in believers' lives prior to the rapture. This is not necessarily unusual, because the man who heavily influenced the rapture "imminency" theory by his pre-tribulation rapture teaching was someone who himself personally opposed the modern manifestation of the supernatural I Corinthians 12 sign gifts. The rapture "imminency" theory is hostile to the supernatural gift of prophecy, because John Nelson Darby, who heavily influenced the rapture "imminency" idea with his pre-tribulational rapture theory, was also hostile to the modern manifestation of the sign gifts (Lindsey, 1999, p. 132).

The formula known as "imminency" is evidently an attempt to give a detailed explanation to the Biblical idea of "watchfulness." The committee formulating the definition for "imminency" was apparently much more than simply successful in their efforts, because endorsement of "imminency" has become a major requirement among many diverse religious organizations today for active membership and participation in those groups adopting it. Unfortunately, some Charismatic and Pentecostal organizations which subscribe to modern manifestations of the I Corinthians 12 sign gifts have adopted the "imminency" formula without apparently realizing that at its most basic theoretical and fundamental level, rapture "imminency" is a formula which is hostile to the gift of prophecy as manifested in predicting future intervening events in the lives of believers which must be fulfilled before the rapture. This kind of manifestation of the gift of prophecy was demonstrated in Biblical times, and is often demonstrated today in the same manner.

There existed specific events in the early church, as seen from the Scriptures, which had to be fulfilled before the rapture. Could it be that there are today intervening events which also must be fulfilled before the rapture? Can the gift of prophecy be manifested today, as in Biblical times, to predict future intervening events in the lives of believers which must be fulfilled before the rapture? It would appear that as in Biblical times, the same gift of prophecy is in operation today. Can those predicted intervening events announce a postponement of the rapture, as they did in the New Covenant Scriptures? If those predictions are from the Holy Spirit, it would appear that they would indeed indicate a delay for the rapture.

The claim is made that the teaching of an *any-moment, imminent rapture requiring no intervening events* promotes watchfulness. Actually, from practical observation, the teaching appears to have the very opposite effect to its apparently claimed purpose. The theory of rapture "imminency," rather than promoting watchfulness, actually appears to have the opposite effect. The theory actually appears to promote apathy. The spreading conventional wisdom which is observable among those who adopt the "imminency" theory is: just be ready. Is readiness equated with "watching?"

What does being ready have to do with watching? Everyone should always be ready, because most believers may not know when they will leave this earth by way of the undertaker. Only an apparently small minority of believers in the panorama of history will leave earth by way of the upper-taker. We should all be ready to leave this world by way of the undertaker at any-moment. Who knows when an aneurism, a stroke, a heart attack, an auto accident, or some other unanticipated circumstance might send a believer into eternity? Who knows whether a plane crash, or some other disaster, could claim some believer from this world in an unexpected instant? Unfortunately, just being ready doesn't fit the Biblical exhortation which Jesus gave to His followers and, by direct and explicit inference, to all believers through all ages down to the present time as indicated in the following Scripture.

37 And what I say unto you I say unto all, Watch. (Mark 13:37)

The doctrine of an *any-moment, imminent rapture requiring no intervening events* actually discourages and undermines the exhortation given by Jesus to "watch." How can you watch for something which could happen at any-moment? How can you watch for something which doesn't have any signposts (intervening events)? What many Biblical interpreters may fail to do is to put the word "watch" into its proper context. Isolating the command to "watch" from its context may lead to any number of erroneous ideas or concepts as to what the command was intended to actually mean. That the exhortation to "watch" is universally applicable to all believers since the disciples is the explicit inference of Jesus Himself. The command was not given only for some selective post-rapture Israelites stuck on earth during the last seven years of the current Gentile era, as some might propose.

The Family Feud for Imminency
I have actually been told by people that they try to avoid thinking about signs which may be linked to end-time events. "I try to avoid thinking about things like that. I just try to be ready." This attitude does not appear to be an isolated or rare instance. It actually appears that this is the currently accepted conventional wisdom which has been widely disseminated and appropriated by a growing number of people. The primary culprit for this attitude is most likely the "imminent rapture" theory. It is appropriate to examine the Biblical basis for "imminency," since that is the issue of relevance to the whole concept of "watching."

If the Bible teaches an *any-moment, imminent rapture requiring no intervening events* for it to occur, where is it found in Scripture? This seems to be the real issue which divides any-moment, imminent rapture teachers from each other.

What may be surprising news to some, but what is exasperating reality to others, is the strange dilemma which exists among imminency rapture pre-tribulational teachers. They are feuding over which verses in the Bible can be used to teach an *any-moment, imminent rapture requiring no intervening events*. There is a lack of agreement and actual disharmony and division over which Bible verses provide a legitimate foundation for the very questionable, but presumably orthodox, doctrinal teaching of an imminent, signless rapture. This issue should not be considered of irrelevance for one very practical reason: many churches and religious organizations require a belief in an *any-moment, imminent rapture requiring no intervening events* as a requirement for membership. Rejecting the formula is grounds for expulsion and excommunication by many of those organizations.

Why is the questionable teaching of "imminency" a standard requirement for membership among so many religious organizations and churches? How can a teaching with such controversial Biblical support become one of the major requirements for actual church membership across such a broad spectrum of groups if it isn't true? The unanimity which appears to exist on the issue seems to lend it great credibility.

One only needs to perhaps recall the historical events associated with the embarrassing spectacles resulting from William Miller's movement, to understand why "imminency" is so universally accepted today. The doctrinal teaching of an any-moment, imminent rapture was probably hailed as an apparently Biblical solution which would prevent a reenactment of the deplorable humiliations associated with the Millerite debacle. The imminency rapture teaching requiring no intervening events may have developed into a formula which was widely adopted because it appeared to be a Biblical concept, and it apparently provided a Biblical reason and method for deterring future, false conclusions like William Miller's. The "imminency" theory probably appeared to be a Biblical solution aimed at ending the practice of setting dates for the rapture.

The "imminency" formula had the appearance of being completely Biblical, and it probably harmonized with the newly circulating theory of a pre-tribulational rapture which had become so influential and appealing in the wake of the Millerite movement. Darby's teachings caught on in America after Miller had defamed the Biblical passages concerning the rapture. Since those days, developments have been taking place among dispensationalists and pre-tribulational rapture groups which have caused division over the issue concerning the Biblical basis for imminency.

The Biblical Basis for Imminency
Several diverse movements exist today among promoters of the *any-moment, imminent rapture requiring no intervening events*. These groups have splintered and divided and developed a variety of theories about the chronology for predicted events in the Scriptures. Some of these groups have developed varying ideas on how the events in the book of Revelation are to be arranged. Not only do they have wide ranging disagreements on the chronology of events, they have differing ideas as to which Biblical verses provide valid support for an "imminent" rapture. Many people might not be aware of these radically different groups, because many of them are in total agreement about one major issue: an *any-moment, imminent rapture requiring no intervening events preventing it can happen at any time*. Such widespread unanimity on this single issue tends to lend an atmosphere of credibility to the whole idea of imminency. After all, the majority can't be wrong. Or can they?

Before the Biblical teaching that the Messiah would return to earth for 1,000 years was recaptured in modern times by use of a literal method for interpreting the prophecies of the Bible, the prevailing teaching in Christian churches was postmillenialism. Today, while there is a limited revival in postmillenial thought, the majority of churches reject it as unbiblical. Among evangelicals today, perhaps the prevailing view is the premillenial theory, which states that Jesus returns to earth to set up His Kingdom for 1,000 years at the end of the present age. This was the belief of the early church and of Biblical Judaism, and it was recaptured in modern times by applying a literal interpretation to Biblically predicted prophetic events (Lindsey, 1999, p. 105).

Just as postmillenialism was at one time the majority opinion, and is now rejected by the majority, so also imminency can be wrong. Might does not make right and the majority can be wrong. It is important to realize that the different groups today who are in agreement over the issue of "imminency" are divided and in disagreement over which Bible passages can legitimately be used to support that imminency. This situation indicates that the perception of unanimity concerning the imminency idea is somewhat superficial.

Probably the major disagreement today among imminency advocates is over which are the proper Biblical verses to be used for teaching imminency. This concern is over whether the gospel passages given by Jesus can be used to teach an imminent rapture. Maybe the majority of traditional dispensationalists reject any of the gospel passages as applying to an imminent rapture. To admit that Jesus in the gospels made statements which directly refer to the rapture opens a Pandora's box which allows the possibility that the rapture can be found in the Messiah's list of end-time events for His disciples. If Jesus taught about the rapture, then maybe the rapture can be found in Messiah's list of events in the Olivet Discourse. To allow that possibility is to allow the possibility of a non-pre-tribulational, non-imminent rapture. For these reasons, probably, most traditional dispensationalists reject the idea of a rapture taught by Jesus in the gospels. This denial of a gospel rapture teaching is maintained by asserting that verses which seem to be about the rapture in the Olivet Discourse are really about the Messiah's return to earth on a white horse, as in the book of Revelation, or to the Mount of Olives, as in Zechariah.

Despite the can of worms which develops when an imminent, any-moment rapture promoter allows a rapture teaching to be found in the words of Jesus, there are some pre-tribulational Bible interpreters who will admit that Jesus did speak about the rapture. Despite the fact they believe in a pre-tribulational, any-moment rapture, some dispensational Bible students will agree that Jesus not only taught about the rapture escape, they will even go so far as to suggest that Jesus Himself laid the primary Biblical foundation for the *imminent rapture which requires no intervening signs.*

What words did Jesus speak which are now causing such controversy? What passages did Jesus teach which have been claimed as Biblical evidence for a rapture escape which could happen at any-moment, maybe even in the next second or two? Let us examine several gospel accounts of those words spoken by Jesus to establish a solid understanding of them.

In Matthew's version of the Olivet Discourse we find a verse which is frequently singled out as supporting an imminent, any-moment rapture which doesn't require any signs.

36 But of that day and hour knoweth no man, no, not

the angels of heaven, but my Father only.
(Matthew 24:36)

Additional words of Jesus which are also identified by some as referring to the "imminent" rapture include the following verses from Matthew's version of the Olivet Discourse.

42 Watch therefore: for ye know not what hour your Lord doth come.

43 But know this, that if the goodman of the house had known in what watch the thief would come, he would have watched, and would not have suffered his house to be broken up.

44 Therefore be ye also ready: for in such an hour as ye think not the Son of man cometh.
(Matthew 24:42-44)

It must be acknowledged that Messiah in these passages does introduce the factors of ignorance, uncertainty, watchfulness, and readiness.

In Mark's version of the Olivet Discourse we find parallel passages to those in Matthew which confirm that Jesus spoke those words.

32 But of that day and that hour knoweth no man, no, not the angels which are in heaven, neither the Son, but the Father.

33 Take ye heed, watch and pray: for ye know not when the time is.

> *34 For the Son of Man is as a man taking a far journey, who left his house, and gave authority to his servants, and to every man his work, and commanded the porter to watch.*
>
> *35 Watch ye therefore: for ye know not when the master of the house cometh, at even, or at midnight, or at the cockcrowing, or in the morning:*
>
> *36 Lest coming suddenly he find you sleeping.*
>
> *37 And what I say unto you I say unto all, Watch. (Mark 13:32-37)*

Again, it must be admitted that Jesus speaks of ignorance, uncertainty, watchfulness, and readiness in these passages which some believe teach imminency.

In Luke's gospel we find a similar account in his version of the Olivet Discourse confirming these words were spoken by Him.

> *34 And take heed to yourselves, lest at any time your hearts be overcharged with surfeiting, and drunkenness, and cares of this life, and so that day come upon you unawares.*
>
> *35 For as a snare shall it come on all them that dwell on the face of the whole earth.*
>
> *36 Watch ye therefore, and pray always, that ye may be accounted worthy to escape all these things that shall come to pass, and to stand before the Son of man. (Luke 2134-36)*

Again, it should be noted that the factors of ignorance, uncertainty, watchfulness, and readiness are all involved in these passages which some use to support the teaching of an *any-moment rapture requiring no intervening events to prevent it.*

Do these words of Jesus apply to the rapture? Some pre-tribulational rapture supporters believe they do. I believe that they do. Perhaps the majority of traditional dispensationalists deny that these passages apply to the rapture, and state they instead refer to the Messianic return to earth when Messiah sets up His visible, thousand year Kingdom on earth. This denial that these passages apply to the rapture by many dispensationalists, has been refuted with some of the most powerful and simple reasoning I have seen. That reasoning was stated by one of dispensationalism's own pre-tribulational teachers (Hunt, 1993, p. 241). His reasoning is that if these verses spoken by Jesus refer to His own return to earth after the seven years of turmoil predicted in the book of Revelation, why would Israel ever be taken by surprise? He argues that the antichrist himself will be expecting Messiah's descent to earth to dispose of his own evil and blasphemous rule. The proposed scenario by those wishing to reject these words by Jesus as speaking about the rapture fails to account for the surprise and uncertainty involved in Messiah's return to earth which do not fit his coming at the end of the seven years of turmoil in the book of Revelation.

Reason and logic suggest that these words of Jesus in the gospel accounts of the Olivet Discourse do apply to the rapture. From these passages have been derived some of the most powerful and persuasive reasons for sign-less rapture imminency ever made by anyone. I contend that if these verses (and similar words by Jesus) do not teach an imminent rapture, there are none anywhere else in the New Covenant which can be validly used to teach an imminent rapture. All other references could only be appealed to circumstantially.

Scripture Interpreting Scripture
The passages from the gospels quoted just previously from the Olivet Discourse by Jesus do indeed refer to the rapture. Not only do they refer to the rapture, those passages also appear to teach an *any-moment, imminent rapture requiring no intervening events.* Major elements of imminency admittedly appear to be involved: ignorance, uncertainty, watchfulness, and readiness.

First there is ignorance. None of the Messiah's disciples know when the Lord will come to rescue His followers (Matt. 24:36, 42, 44; Mark 13:32, 33, 35-36; Luke 21:34, 36).

Secondly, there is uncertainty. The Messiah comes when apparent followers are not expecting it (Matt. 24:42, 44; Mark 13:33, 35, 36, 37; Luke 21:34).

Thirdly, watchfulness is required (Matt. 24:42, 43; Mark 13:33, 35, 37; Luke 21:35).

Fourthly, readiness is required ((Matt. 24:44; Mark 13:33, 35-36; Luke 21:34, 36).

All these factors seem convincingly to prove an imminent rapture requiring no intervening events is being taught by Jesus. There is just one little problem. When Scripture interprets Scripture, authoritative teaching takes place, and other Scriptures undermine the premature and superficial conclusion that these passages teach an *imminent rapture requiring no intervening events.*

If the previously quoted passages from the Olivet Discourse can be shown and demonstrated not to teach an any-moment, imminent rapture requiring no intervening events, there aren't any Scriptures anywhere in the Bible which can be used to prove an unconditionally imminent and any-moment rapture. The attempt could be made, but all other passages could at best be only appealed to in a circumstantial manner to support an imminent rapture.

In examining the four conditions which seem to teach a sign-less and imminent rapture, we have considered ignorance of the event, uncertainty about the timing of the event, watchfulness which is required for the Messiah's future rescue, and readiness for that rescue. There is an element which has not been considered about these four items. Some of these items are conditional, because Jesus Himself confirmed that the rapture event cannot happen at any-moment in one of the very passages we have examined which supposedly teaches imminency. In Matthew's version of the Olivet Discourse, Jesus assures His disciples that the rapture rescue cannot happen at any-moment.

> *36 But of that day and hour knoweth no man, no, not the angels of heaven, but my Father only.*
> *(Matthew 24:36)*

A close inspection of this verse contrasts the ignorance of everyone else about the rapture to the specific knowledge which exists in the divine heavenly Father. Jesus was absolutely certain His Father knew the exact day and hour at which the rapture would occur. If the Father knows the specific day and specific hour when the rapture will occur, it cannot obviously happen at any-moment.

"Aha!" Some will immediately state, "The rapture can't

happen at any-moment from the divine point of view, but from a human point of view, the rapture could take place at any-moment."

Not really. Jesus even negates this concept. Jesus demonstrates from the very passages which we have already examined, as well as from others, that certain conditions exist for uncertainty to be involved. True, Jesus said believers do not know the exact day and hour at which the rapture will occur far in advance of that event, but Jesus specifically denies that the rapture will unconditionally occur at any-moment without any signs.

In Luke's version of the Olivet Discourse, we observe a condition which Jesus specifically indicates must exist in order for a follower to be caught by surprise at the rapture.

> *34 And take heed to yourselves, lest at any time your hearts be overcharged with surfeiting, and drunkenness, and cares of this life, and so that day come upon you unawares. (Luke 21:34)*

Notice the condition which must exist for a follower of Messiah's to be caught by surprise. That condition is distraction. Only if a person is distracted from watching for the Lord's coming will the follower be caught by surprise. The analogy is so very characteristic of everyday life it almost doesn't need an explanation, but one will be offered. Suppose you have been contacted by someone who states that they will arrive in twenty minutes. You start to prepare for your visitor, but then the phone rings. You begin talking to someone you haven't seen for years, and you suddenly are involved in a conversation which causes you to lose track of the time and the coming of your expected visitor. When your visitor arrives, you are not ready and you are caught by surprise. You were distracted from watching. If you had not been distracted, you would have been ready to go and would have prepared. You could then have watched for their vehicle, and you could have been out on the sidewalk by the time they arrived.

In the subsequent verse to Luke 21:34 Jesus infers that this is the whole reason non-believers are taken by surprise, they are not watching or looking for the coming of the Messiah.

> *34 And take heed to yourselves, lest at any time your hearts be overcharged with surfeiting, and drunkenness, and cares of this life, and so that day come upon you unawares.*
>
> *35 For as a snare shall it come on all them that dwell on the face of the whole earth.*
>
> *36 Watch ye therefore, and pray always, that ye may be accounted worthy to escape all these things that shall come to pass, and to stand before the Son of man. (Luke 21:34-36)*

Two verses in this passage warn against being distracted (Luke 21:34-35). The danger of distraction forms the very foundation for the warning to watchfulness in verse thirty-six. Unbelievers are not watching at all, and some believers may become distracted from watching for the Lord's coming.

Someone may claim these verses have nothing to do with signs, but only with believers knowing that the Lord is coming, and that we are to be ready. So let us examine an explanatory illustration given by Jesus which applies to this whole concept of the rapture.

> 43 But know this, that if the goodman of the house had known in what watch the thief would come, he would have watched, and would not have suffered his house to be broken up.
> (Matthew 24:43)

Here Jesus uses a metaphor which compares His rapture rescue of followers to that of a thief. Jesus stated that watching would have prevented the theft from taking place. The idea is that if someone is not watching, that individual will experience Messiah's coming as similar to that of a thief (believers will have been stolen from earth). If, on the other hand, one is watching, the experience will not be like a burglary (they will be rescued from earth by the Messiah).

Jesus confirms His own thief metaphor in another place, where He again refers to His coming as a thief. In this other passage, Jesus made some explanatory comments which explained His meaning a little more clearly. That passage specifically indicates that signs are involved in the "watch." This is clearly implied because He indicates almost the exact hour of His coming can be known by watching.

> 3 ... If therefore thou shalt not watch, I will come on thee as a thief, and thou shalt not know what hour I will come upon thee. (Revelation 3:3)

Here Jesus specifically explains that watching *prevents* a "theft" experience, but not watching *causes* a thief experience. This passage reinforces the thief metaphor in the Olivet Discourse in very explicit terms. The implication is,

The Planetary Exodus

if one is watching, they may know the very *hour* at which Jesus will come. If a person is distracted from watching for the Messiah's rescue, the experience they have of the coming of Messiah will be like that of a thief, and the distracted person won't know *even the approximate hour* at which the Lord will come. Watching causes one to be ready to be rescued, so watching enables a person to experience the rescue. Not watching causes one to be unprepared, so not watching means the person is left on earth when believers are snatched away. The person who failed to watch experiences a burglary when those who did watch are stolen from earth. This experience is very similar to watching for a visitor in everyday life.

That the thief metaphor applies to the rapture is virtually certain from a similar passage in Paul's first letter to the Thessalonians. In that particular passage of Scripture the Apostle Paul explains the thief metaphor and elaborates on the meaning. This passage is generally regarded by most pre-tribulational promoters as a reference to the rapture.

> *2   For yourselves know perfectly that the day of the Lord so cometh as a thief in the night.*

> *3   For when they shall say, Peace and safety; then sudden destruction cometh upon them, as travail upon a woman with child; and they shall not escape.*

> *4   But ye, brethren, are not in darkness, that that day should overtake you as a thief.*

> *5   Ye are all the children of light, and the children of the day: we are not of the night, nor of darkness.*

> *6   Therefore let us not sleep, as do others; but let us*

watch and be sober.

7 For they that sleep sleep in the night; and they that be drunken are drunken in the night.

8 But let us, who are of the day, be sober, putting on the breastplate of faith and love; and for an helmet, the hope of salvation. (I Thessalonians 5:2-8)

The Apostle Paul confirms that the rapture rescue is a thief-like experience for non-watchers, and non-watchers are not rescued. The Apostle contrasts the thief-like experience which non-believers have when they are left behind as believers are stolen from earth, to the rescue experience watchers have at the rapture. Someone who is not watching is described as a sleeper of the night, a child of darkness. Someone who watches is a child of light, a child of the day. That signs are involved is implied by the statement in verse four where the Apostle states, "But ye, brethren, are not in darkness, that that day should overtake you as a thief."

The Apostle from the tribe of Benjamin infers that the uncertain time for the rapture is only uncertain for those who are not watching (the children of the night who are asleep), but for children of the day the clear inference is that the timing of the rapture will not be surprising at all, because children of the day will have been watching and so will be aware of the closeness of that event.

Watchfulness, it appears, allows believers to stay informed about the proximity, the closeness or farness, of the rapture rescue event, as the writer of Hebrews indicates.

25 Not forsaking the (earthly) assembling of ourselves together, as the manner of some is; but exhorting one another: and so much the more, as ye see the day (of heavenly assembling) approaching. (Hebrews 10:25)

To paraphrase the apparent pun from the Greek, we can interpret this verse to read as follows. "Have more earthly synagogue meetings, as you see the day of the heavenly synagogue meeting approaching."

The thief metaphor corresponds to what the writer of Hebrews seems to suggest. The day of the rapture can be observed as a day which approaches, so that the nearness or farness of the day of the rapture can be visually measured. The distance of the day of the rapture can be visually measured. How does one measure the distance of something visually? They use signs. Sign language is a required skill for determining the distance of the time for the rapture. William Miller had not learned the appropriate sign language.

Developing Sign Language Skills
An acquaintance had phoned me indicating their arrival in about fifteen minutes. I did not know whether they drove a car, van, or truck. While waiting for him I focused on the behavior of vehicles as the only method for determining when he had arrived. A slowing vehicle would be a sign my visitor was near. The only vehicle I had seen which slowed down and appeared to get ready to stop was a white car. The car actually slowed down while it was on the opposite side of the street. That was the first evidence I had concerning the arrival of my expected visitor. The car then drove off, and evidently made a U-turn, and was out of sight for a few moments until it came back and the vehicle reappeared on my side of the street. I couldn't see the driver's face at all when he was on my side of the street. Without knowing absolutely who it was, I had come to the conclusion, only

by the observable signs I had seen, that this was my expected visitor. No-one can claim I was not looking for my expected visitor just because I was focused on the behavior of a car. Sometimes signs are the only practical means by which a person can watch for someone who is expected in everyday life. Signs help to determine the nearness or farness of the expected visitor's arrival.

Watching must include signs. This is so practical in everyday life. Signs are also implied by the context of Messiah's exhortation to "watch." The teaching of an *any-moment, imminent rapture requiring no intervening events* actually discourages and undermines the exhortation given by Jesus to "watch," because there are no signs to watch for from that point of view. What many Biblical interpreters may fail to do is to put the word "watch" into its proper context. Isolating the command to "watch" from its context may lead to any number of erroneous ideas or concepts regarding what the command actually infers.

To understand what "watch" meant in the Olivet Discourse, we need to place it back into its context. The immediate context indicates watching for the Messiah's coming, but the larger context concerns the entire Olivet Discourse. To understand how Messiah's command for everyone to "watch" was intended to be understood, we must also take the larger context of the Olivet Discourse and examine how that dialogue began to develop.

> 3 And as he sat upon the mount of Olives, the disciples came unto him privately, saying, Tell us, when shall these things be? and what shall be the sign of thy coming, and of the end of the world [age]? (Matthew 24:3)

The Olivet Discourse began when the disciples heard Jesus predict the destruction of Israel's national religious monument, the sacrificial Temple. The first questions which popped into their minds were the following. 1) When is that going to happen? 2) What will the signs be for your coming? 3) What are the signs for the end of the age?

Did Jesus avoid their questions? No. Jesus answered all three questions in the extended comments we call the Olivet Discourse. Jesus gave sign after sign after sign. Were these signs for believers to watch? If not, why were they given? Evidently, when Jesus finished answering the disciples' questions, He more or less made the following statement.

> 37 "Hey! You fellows wanted signs. I have just given you a whole series of signs to think about. Now let me give you a word of advice. Watch. Watch for the signs you men have just asked me about. Watch for the signs I have just given you. The same goes for everyone else. Watch."
> (Author's interpretational paraphrase of Mark 13:37)

Sign language is a required skill for determining the distance of the time for the rapture. William Miller had not learned the appropriate sign language. The signs in the Olivet Discourse were given as proximity gauges to enable the nearness or farness of events to be assessed and evaluated by the disciples. Not only did Jesus give these signs to the disciples as events to be watched for, these signs were for all subsequent followers of the Messiah throughout all succeeding centuries.

> 37 And what I say unto you I say unto all, Watch.
> (Mark 13:37)

It appears that the closer a watching follower of the Messiah gets to the rapture event, the more precise their ability becomes to accurately identify the actual time for that event.

> 3 ... *If therefore thou shalt not watch, I will come on thee as a thief, and thou shalt not know what hour I will come upon thee. (Revelation 3:3)*

Jesus clearly implied in this Scripture that watching might enable a believer to know almost the exact *hour* of His coming. Sign language is a required skill for determining the distance of the time for the rapture. William Miller had not learned the appropriate sign language.

How Close is the Rapture?

Jesus gave an unmistakable sign signaling when the rapture would be soon to occur. That sign is the event which reveals the actual identity of the future tyrant of earth. An unmistakable sign for the rapture is the fulfillment of an event which will identify who the false Messiah is, as has previously been discussed in other chapters. That sign was given by both the Apostle Paul, and Jesus.

Conclusion

What are the implications of the ideas arrived at in this chapter? Perhaps the clearest implication is that one must learn the appropriate sign language. Another implication is that the use of signs for gauging the nearness or farness of the rapture event can be, and most likely will be, abused. Jesus Himself gave the clearest indication that this abuse would occur in the context of passages some have used to formulate the erroneous idea of an *imminent rapture which can happen at any-moment without any intervening events to prevent it from happening.* That formula denies that there is a sign language. Sign language is a required skill for determining the nearness or farness for the time for the rapture. William Miller had not learned the appropriate sign language. Rapture "imminency" virtually denies that a sign language exists. No sign language is almost as bad as the wrong sign language.

How will the Biblical sign language authored by Jesus and the Apostle Paul be abused? The following verses indicate how this abuse will occur.

> *48 But and if that evil servant shall say in his heart, My lord delayeth his coming;*
>
> *49 And shall begin to smite his fellowservants, and to eat and drink with the drunken;*
>
> *50 The lord of that servant shall come in a day when he looketh not for him, and in an hour that he is not aware of,*
>
> *51 And shall cut him asunder, and appoint him his portion with the hypocrites: there shall be weeping and gnashing of teeth. (Matthew 24:48-51)*

These passages indicate that a false sense of security at one's estimation for the time of the coming of the Lord may lead some to distraction and unreadiness for that event. In these passages Jesus indicates that misuse of the signs may lead some into apostate life-styles resulting in eternal alienation from the Lord and unreadiness for His rescue.

The Apostle Paul infers that the uncertain time for the rapture is only uncertain for those not watching (the children of the night), but for children of the day the time of the rapture will not be surprising because children of the day will, by watching, be aware of the closeness of that event.

Watchfulness, it appears, allows believers to stay informed about the proximity, the closeness or farness, of the rapture rescue event.

Signs in the Olivet Discourse were given as proximity gauges to enable the nearness or farness of events to be assessed by the disciples of Jesus, and by all subsequent believers throughout all subsequent centuries since then. Jesus stated this after He had answered all of the disciples' questions about signs in the Olivet Discourse.

> *37 "Hey! You fellows wanted signs. I have just given you a whole series of signs to think about. Now let me give you a word of advice. Watch. Watch for the signs you men have just asked me about. Watch for the signs I have just given you. The same goes for everyone else. Watch."*
> *(Author's interpretational paraphrase of Mark 13:37)*

A Word

Perhaps you are Jewish, and you are saying to yourself, "I could never be Jewish and accept Jesus as my Lord. I would no longer be Jewish."

What if Jesus, Yeshua, is the real Messiah of Israel? Can you really be a Jew in the truest sense if you reject the real Jewish Messiah? The answer is, "No." A Jew in the truest sense will not reject the real Messiah. If Jesus, Yeshua, is the real Messiah of Israel, you cannot be a Jew in the most important sense if you reject Him as your Messiah, no matter what your biological lineage may be. Being Jewish in the truest sense consists of circumcision of the heart, which is what makes a person a spiritual Jew (Leviticus 26:41, Romans 2:25-29). Biological lineage is a superficial issue compared to the real affairs of the soul and spirit.

If you are a Jew or a Gentile, you can enter into eternal life by partaking of the person who is the passover sacrifice for sins. A zealous Jewish Pharisee of the tribe of Benjamin, the Apostle Paul, stated in I Corinthians 5:7 that, "Christ (Messiah) our passover is sacrificed for us."

If you don't know Israel's Messiah as your Lord and Savior, why not invite Him into your life? He died for us.

According to II Corinthians 5:7, "For he hath made him to be sin for us, who knew no sin, that we might be made the righteousness of God in him."

The prophet Isaiah wrote of Messiah and stated the following words.

> *5 But he was wounded for our transgressions, he was bruised for our iniquities: the chastisement of our peace was upon him; and with his stripes we are healed.*
>
> *6 All we like sheep have gone astray; we have turned every one to his own way; and the LORD hath laid on him the iniquity of us all. (Isaiah 53:5-6)*

Jesus, Yeshua the Messiah, asks everyone to follow Him.

> *28 Come unto me, all ye that labour and are heavy laden, and I will give you rest.*
>
> *29 Take my yoke upon you, and learn of me; for I am meek and lowly in heart: and ye shall find rest unto your souls.*
>
> *30 For my yoke is easy, and my burden is light. (Matthew 11:28-30)*

If you really mean business with the Lord of Israel, you have His guarantee.

> *13 And ye shall seek me, and find me, when ye shall search for me with all your heart. (Jeremiah 29:13)*

You can seek the Lord today and ask the God of Israel to

cleanse you through Jesus, Yeshua the Messiah, and then ask Him to help you to begin following Him.

How can you follow the Messiah? Read His words. Find out His plans for your life directly from Him. Meet with others who know the Messiah, and share your spiritual knowledge about Him with others.

References

Alford, Henry, *Alford's Greek Testament, An Exegetical and Critical Commentary.* Grand Rapids, Michigan: Guardian Press, printed 1976.

Aquinas, Thomas (Various Translators), *Catena Aurea.* Albany, New York: Preserving Christian Publications, 2000.

Beale, G.K., *The New International Greek Testament Commentary, The Book of Revelation.* Grand Rapids, Michigan: William B. Eerdmanns Pub. Co., 1999.

Beaseley-Murray, George R., *Jesus and the Last Days, The Interpretation of the Olivet Discourse.* Peabody, Massachussetts: Hendrikson Publishers, 1993.

Chandler, Russell, *Doomsday, The End of the World–A View Through Time.* Ann Arbor, Michigan: Servant Publications, 1993.

Charles, R.H., *A Critical and Exegetical Commentary on The Revelation of St. John.* Edinburgh: T. & T. Clark, printed 1985.

Clarke, Adam, *The Holy Bible Containing the Old and New Testaments with A Commentary and Critical Notes*. Nashville: Abingdon, n.d.

Cohen, A., *Soncino Books of the Bible, The Twelve Prophets*. London/New York: The Soncino Press, printed 1985.

Dake, Finis Jennings, *Revelation Expounded, or Eternal Mysteries Simplified*. Lawrenceville, Georgia: Dake Bible Sales, Inc., 1977.

Eadie, John, *Greek Text Commentaries, Thessalonians*. Grand Rapids, Michigan: Baker Book House, Reprinted 1979.

Gundry, Robert H., *The Church and the Tribulation*. Grand Rapids, Michigan: Zondervan Pub. House, 1973.

Henry, Matthew, *A Commentary on the Holy Bible*. New York: Funk & Wagnalls, n.d.

Hunt, Dave, *When Will Jesus Come?* Eugene, Oregon: Harvest House Publishers, 1993.

(JFB) Jameison, Robert; Fausset, A.R.; Brown, David; *A Commentary Critical and Explanatory on the Whole Bible*. Hartford, Conn.: S.S. Scranton Co., n.d.

Ladd, George Eldon, *A Commentary on the Revelation of John*. Grand Rapids, Michigan: William B. Eerdmanns Pub. Co., 1972.

Lange, John Peter, *A Commentary on the Holy Scriptures: Critical, Doctrinal, and Homiletical, with Special Reference to Ministers and Students*. New York: Charles Scribner, & Co., 1867.

Lindsey, Hal, *Vanished into Thin Air.* Beverly Hills, CA: Western Front, Ltd., 1999.

MacPherson, Dave, *The Incredible Cover-Up.* Plainfield, New Jersey: Logos International, 1975.

McDowell, Josh, *Evidence That Demands A Verdict, Revised. Ed.* San Bernadino, CA: Here's Life Publishers, Inc. 1979.

Moore, Philip N., *A Liberal Interpretation on the Prophecy of Israel–Disproved.* Atlanta, Georgia: The Conspiracy Inc., 1997.

Mueller, R., Compiler; Anstadt P., Translator; *Luther's Explanatory Notes on the Gospels.* York, PA.: P. Anstadt & Sons, 1899.

Peters, George N.H., *The Theocratic Kingdom,* Vol. 3. Grand Rapids, Michigan: Kregel Publications, printing 1994.

Pink, Arthur W., *The Antichrist.* Grand Rapids, Michigan: Kregel Publications, printed 1988.

Pringle, William, translator, *Commentary on a Harmony of the Evangelists, Matthew, Mark, Luke by John Calvin.* Grand Rapids, Michigan: Wm. B. Eerdmanns Pub. Co., printed 1949.

Schaff, Philip; Wace, Henry; *A Select Library of Nicene and Post-Nicene Fathers of the Christian Church, Second Series.* New York: The Christian Literature Company, 1896.

Spence, H.D.M.; Excell, Joseph S.; Editors; *The Pulpit Commentary.* Grand Rapids, Michigan: Wm. B. Eerdmanns Pub. Co., Reprinted 1961.

Van Impe, Jack, *Revelation Revealed.* Troy, Michigan: Jack Van Impe Ministries, 1982.

Van Kampen, Robert, *The Rapture Question Answered, Plain & Simple.* Grand Rapids, Michigan: Fleming H. Revell, 1997.

Walvoord, John F., *The Prophecy Knowledge Handbook.* U.S./Canada/England: Victor Books, 1990.

Walvoord, John F; Zuch, Roy B.; Editors; *The Bible Knowledge Commentary, An Exposition of the Scriptures by Dallas Seminary Faculty.* Wheaton, Illinois: Victor Books, 1983.

Woods, Dennis James, *Unlocking the Door: A Key to Biblical Prophecy.* Lafayette, Louisiana: Huntington House, 1994.

Young, Robert, *Analytical Concordance to the Bible.* Peabody, Massachusetts: Hendrikson Publishers, n.d.

www.ingramcontent.com/pod-product-compliance
Lightning Source LLC
Chambersburg PA
CBHW020723180526
45163CB00001B/90